AMTE Monograph Series Volume 5

Inquiry into Mathematics Teacher Education

Edited by

Fran Arbaugh
University of Missouri

P. Mark Taylor
Koinonia Associates

Monograph Series Editor

Denisse R. Thompson
University of South Florida

Published by

Association of Mathematics Teacher Educators
San Diego State University
c/o Center for Research in Mathematics and Science Education
6475 Alvarado Road, Suite 206
San Diego, CA 92129

www.amte.net

Library of Congress Cataloging-in-Publication Data

Inquiry into mathematics teacher education / edited by Fran Arbaugh, P. Mark Taylor.
p. cm. -- (AMTE monograph series ; v. 5)

ISBN 978-1-932793-06-2

1. Mathematics teachers--Training of--United States. 2. Mathematics--Vocational guidance--United States. I. Arbaugh, Fran. II. Taylor, P. Mark. III. Association of Mathematics Teacher Educators.

QA10.5.I57 2008
510.71--dc22

2008055023

Printed in the United States of America

Contents

Foreword vii
Jennifer Bay-Williams, University of Louisville

1. **Inquiring into Mathematics Teacher Education** 1

Fran Arbaugh, University of Missouri
P. Mark Taylor, Koinonia Associates

2. **Linking Master Teachers and Mathematics Educators to Build a Statewide Professional Development Academy** 11

Terry Goodman, University of Central Missouri
Larry Campbell, Missouri State University

3. **Toward a Situative Perspective on Online Learning: Metro's Mathematics for Rural Schools Program** 23

Brooke Evans, The Metropolitan State College of Denver
Hamilton Bean, University of Colorado at Boulder
Lew Romagnano, The Metropolitan State College of Denver

4. **The Task Design Framework: Considering Multiple Perspectives in an Effective Learning Environment for Elementary Preservice Teachers** 35

Kathryn B. Chval, University of Missouri
John K. Lannin, University of Missouri
Angela D. Bowzer, Westminster College

5. **Concentric Task Sequences: A Model for Advancing Instruction Based on Student Thinking** 47

Laura R. Van Zoest, Western Michigan University
Shari L. Stockero, Michigan Technological University

6. **Using the Problem-Solving Cycle Model of Professional Development to Support Novice Mathematics Instructional Leaders** 59

Karen Koellner, University of Colorado Denver
Craig Schneider, University of Colorado at Boulder
Sarah Roberts, University of Colorado at Boulder
Jennifer Jacobs, University of Colorado at Boulder
Hilda Borko, Stanford University

7. **Case Stories: Supporting Teacher Reflection and Collaboration on the Implementation of Cognitively Challenging Mathematical Tasks** 71

Elizabeth K. Hughes, University of Northern Iowa
Margaret S. Smith, University of Pittsburgh
Melissa Boston, Duquesne University
Michael Hogel, Mt. Lebanon School District

8. **Heeding the Call: History of Mathematics and the Preparation of Secondary Mathematics Teachers** 85

Kathleen M. Clark, Florida State University

9. **Developing Mathematical Pedagogical Knowledge by Evaluating Instructional Materials** 97

Margret A. Hjalmarson, George Mason University
Jennifer M. Suh, George Mason University

10. **Shifting Roles, Shifting Perspectives: Experiencing and Investigating Pedagogy in Teacher Education** 109

Michael D. Steele, Michigan State University

11. **Lesson Study as Professional Development for Mathematics Teacher Educators** 119

Jo A. Cady, University of Tennessee
Theresa M. Hopkins, Tennessee Governor's Academy
Thomas E. Hodges, Tennessee Governor's Academy

12. ***Kyozaikenkyu*: A Critical Step for Conducting Effective Lesson Study and Beyond** 131

Tad Watanabe, Kennesaw State University
Akihiko Takahashi, DePaul University
Makoto Yoshida, William Paterson University

13. **Technological Pedagogical Content Knowledge (TPCK): Preparation of Mathematics Teachers for 21st Century Teaching and Learning** 143

Margaret L. Niess, Oregon State University
Robert N. Ronau, University of Louisville
Shannon O. Driskell, University of Dayton
Olga Kosheleva, University of Texas at El Paso
David Pugalee, University of North Carolina at Charlotte
Marcia Weller Weinhold, Purdue University Calumet

14. **Scholarship for Mathematics Educators: How Does This Count for Promotion and Tenure?** 157

Michelle K. Reed, Wright State University
Susann M. Mathews, Wright State University

Bay-Williams, J.
AMTE Monograph 5
Inquiry into Mathematics Teacher Education
©2008, pp. vii-viii

Foreword

Welcome to the fifth AMTE Monograph: *Inquiry into Mathematics Teacher Education.* The 14 chapters in this monograph provide support for mathematics teacher educators in both their Practical Knowledge, or that knowledge gained through reflection on practice, and their Professional Knowledge, which encompasses practical knowledge and includes research-based knowledge (Hiebert, Gallimore, & Stigler, 2002). (See Arbaugh and Taylor for an elaboration of these types of knowledge as well as more detail on the contents of the monograph.) As Arbaugh and Taylor suggest, individually these articles provide insights into advancing our thinking about professional development, teacher preparation, and program development. Collectively, they have the potential to help the field of mathematics teacher education move forward in framing effective practices in mathematics teacher education and developing a focused, cohesive research agenda. ATME's Monograph 5, therefore, is a superb resource for mathematics teacher educators as they think about their own practice and about engaging in inquiry related to mathematics teacher education. The context in which this monograph has emerged is worth mention.

This year, 2008, brought much attention to mathematics teacher education, in particular to the quality of mathematics teachers' knowledge. In February the National Mathematics Advisory Panel (NMP) released their report, titled *Foundations for Success,* in which they discussed the research findings related to connecting mathematics teachers' content knowledge to student achievement. This panel found that teachers' content knowledge plays an important role in their effectiveness -- most importantly when the teachers' content knowledge is directly related to the mathematics that they teach (as compared to teachers' understanding of advanced content knowledge).

In the summer of 2008 the National Council of Teacher Quality (NCTQ) released its report, titled *No Common Denominator: The Preparation of Elementary Teachers in Mathematics by America's Education Schools.* Through their review of various lists of recommended content for elementary teachers, the report recommends that all teacher preparation programs offer nine hours of mathematics content that is specific to teachers (not general college mathematics courses).

In the fall of 2008 the National Academy of Education (NAE) released a series of White Papers, including one titled "World Class Science and Mathematics." Included in this two-page brief is a recommendation for increased research on effective teaching, recognizing the importance of developing our professional knowledge:

> Teaching science and mathematics is highly skilled, knowledge-intensive, complex work. …In addition to knowing the subject matter, science and

> mathematics teachers must have a specialized kind of knowledge that enables them to engage students in active conceptual learning. (p. 2)

Although these reports (and others like them) differ in their specific findings and recommendations, what they have in common is (1) an awareness of the importance of mathematics teacher education and (2) the need for a stronger professional knowledge in mathematics teacher education.

More and more, the education community looks to AMTE to provide leadership in moving the field forward in the professional knowledge for teaching mathematics. This monograph, as well as the four that precede it, represents one of the important ways that AMTE is contributing to this effort. It is through their commitment to mathematics teacher education that the contributing authors, the editorial panel, and the editors made this important resource possible.

The review process for AMTE Monograph 5 was as follows. The editors received 46 manuscripts for review. Each manuscript underwent a blind review by two editorial board members and one editor. When all reviews were complete, the editors chose 13 manuscripts for inclusion in the monograph, based on numerical review scores and reviewers' comments.

Collectively, these editors and chapter authors contributed to a tremendous resource that provides a breadth and depth of practical and professional knowledge – a resource that can lead us to reflect and engage in inquiry on mathematics teacher education.

Jennifer M. Bay-Williams
AMTE President 2007-2009

Arbaugh, F. and Taylor, P. M.
AMTE Monograph 5
Inquiry into Mathematics Teacher Education
©2008, pp. 1-9

1

Inquiring into Mathematics Teacher Education

Fran Arbaugh
University of Missouri

P. Mark Taylor
Koinonia Associates

The work of mathematics teacher educators has never been more important than in this era of accountability. Mathematics teachers (both preservice and inservice) grapple with ways to support their students in developing mathematical proficiency (as defined by the National Research Council, 2001) in classroom environments where students' learning is focused on sense-making, mathematical authority is shared, and students have the opportunity to learn important mathematics. As researchers have documented (e.g., Sowder, 2007), it is difficult for mathematics teachers to "unlearn" how to teach mathematics (Ball, 1988), given the "apprenticeship of observation" (Lortie, 1975) they have undertaken in their own K-12 mathematics schooling.

Across the United States, mathematics teacher educators (MTEs) have undertaken this charge of working to improve mathematics teaching (which, in turn, will improve K-12 students' opportunities to learn mathematics). Many MTEs implement novel and innovative approaches to mathematics teacher development with preservice and/or inservice teachers, constantly seeking to understand those practices and their impact on mathematics teacher education (MTE) students. In an effort to share what we learn about our practices, MTEs have begun to heed the advice that we give to K-12 mathematics teachers – we too are opening up our classrooms and practices for others to consider. As evidenced by the number of submissions we had for this monograph (45) as well as the increasing number of applications to speak at the AMTE Annual Meetings, our community is responding to the call to share what we are learning through inquiring into our MTE practices.

In this opening chapter of the 5th AMTE Monograph, we argue for the necessity of coordinating our efforts and bringing coherence to our common knowledge. As a community, MTE is relatively "young" when compared to other educational communities. The *Journal of Mathematics Teacher Education* (*JMTE*) was established in 1998; the AMTE constitution was ratified in 1994.

As we continue to seek avenues for coordinating what we are learning into a more coherent whole, building on the work of another community may be a good starting point for our efforts. In the next section, we present work done in the area of knowledge bases for the teaching profession and then present an adaptation of those ideas for the MTE profession.

Practical Knowledge and Professional Knowledge

Hiebert, Gallimore, and Stigler (2002) assert that the knowledge base for the teaching profession consists of two domains: *practical knowledge* and *professional knowledge*. Practical knowledge, as defined by Hiebert, Gallimore, and Stigler, consists of "the kinds of knowledge practitioners generate through active participation and reflection on their own practice" (p. 4). Professional knowledge encompasses practical knowledge and also includes research-based knowledge – knowledge that is based on empirical research studies about teaching. These authors argue that teachers often make instructional decisions based on their practical knowledge and rarely seek out the research literature to inform their teaching, even though the research literature contains findings that could inform their practice. This situation creates a need to find ways to link research and practice more effectively. Hiebert and his colleagues ask the question, "Is there a road that could lead from teachers' classrooms [practical knowledge] to a shared, reliable, professional knowledge base for teaching?" (p. 4).

We contend that the same two domains exist for the knowledge base for mathematics teacher education. We have MTE practical knowledge – knowledge that we build on a daily basis while actively participating and reflecting on our practices as mathematics teacher educators. We often share that knowledge with each other while chatting in our offices (e.g., "Let me tell you what happened in my methods class today!") as well as in venues such as the AMTE Annual Meetings and the AMTE Monograph series. Many of the chapters contained in this monograph could be described as MTEs sharing their practical knowledge with other MTEs.

In addition, in the mathematics teacher education community, we also have knowledge that the research community establishes (see, for example, articles published in the *JMTE*). The studies presented in venues like *JMTE* play an important role in the MTE community, adding to the professional knowledge base for mathematics teacher education. As MTEs we learn from others' practical and professional knowledge. However, we contend that our MTE community does not have a "shared, reliable, and professional knowledge base" (see the Hiebert, Gallimore, and Stigler quote above). A possible direction for the MTE community is to build from the work of one of the authors in this monograph – Hilda Borko. In the next sections, we present a way that we could, as a community, frame our work of inquiring into mathematics teacher education, and begin to coordinate the knowledge base for MTE.

Framing Inquiries into Mathematics Teacher Education

At the 2004 annual meeting of the American Educational Research Association, Hilda Borko focused her Presidential Address on research in teacher professional development. Subsequently, she wrote an article titled "Mapping the Terrain in Research on Professional Development," which appeared in *Educational Researcher* (Borko, 2004). In that article, Borko identified four key elements that comprise any professional development system:

- The professional development program;
- The teachers, who are learners in the system;
- The facilitator, who guides the teachers as they construct new knowledge and practices; and
- The context in which the professional development occurs. (p. 4)

Borko then presented a three-phase framework for considering research on professional development (see Table 1).

Table 1: *Phases of Research on Teacher Professional Development* (Borko, 2004, p. 4)

Phase of Professional Development Research	Context of Research Study	Research Focus
1	Researchers focus on an individual professional development program at a single site.	Researchers typically study the professional development program, teachers as learners, and relationships between these two elements of the system. The facilitator and context remain unstudied.
2	Researchers study a single professional development program enacted by more than one facilitator at more than one site.	Researchers explore the relationships among facilitators, the professional development program, and teachers as learners.
3	Research focus broadens to comparing multiple professional development programs, each enacted at multiple sites.	Researchers study the relationships among all four elements of a professional development system: facilitator, professional development program, teachers as learners, and context.

Borko's description of the elements of a professional development system can also be used to describe preservice mathematics teacher education:

- The mathematics teacher education program;
- The preservice mathematics teachers, who are learners in the system;
- The mathematics teacher educator, who guides the preservice teachers as they develop new knowledge and practices; and
- The context in which the mathematics teacher education program occurs.

Similarly, Borko's phases of research on professional development easily map onto research on preservice mathematics teacher education (see Table 2).

Table 2: *Borko's (2004) Three Phases Adapted for Preservice Mathematics Teacher Education*

Phase of Research on Preservice Mathematics Education	Context of Research Study	Research Focus
1	Researchers focus on an individual preservice mathematics teacher education course or program at a single university or site.	Researchers typically study the preservice mathematics teacher program, preservice teachers as learners, and relationships between these two elements of the system. The mathematics teacher educator and context remain unstudied.
2	Researchers study a single preservice education course or program enacted by more that one mathematics teacher educator at more than one university or site.	Researchers explore the relationships among mathematics teacher educators, the mathematics teacher education program, and preservice teachers as learners.
3	Research focus broadens to comparing multiple preservice teacher programs, each enacted at multiple universities or sites.	Researchers study the relationships among all four elements of a professional development system: mathematics teacher educator, mathematics teacher education program, preservice teachers as learners, and context.

If we are to make progress in coordinating all of our work in MTE (and establishing a deeper, more connected professional knowledge base), then extending Borko's framework for research on professional development to include all of the work of those who are inquiring into mathematics teacher education may be useful.

However, despite the progress made in the whole of mathematics education research over the last fifty years, the vast majority of the work in mathematics teacher education fails to surpass Phase 1, a situation confirmed by the work of the National Mathematics Advisory Panel (2008), which argued that little is empirically documented, and thus known, about the practices of MTEs and the results of those practices. This situation was also confirmed in a study of the literature on mathematics methods courses (Taylor & Ronau, 2006). This trend toward Phase 1 inquiries is also evident in the chapters in this monograph. As important as it is that we share our MTE practical knowledge, we also need to be looking to the future and how we can address our critics.

In the following section, we present the chapters contained in this monograph by making explicit the connections that we see among the authors' inquiries into MTE. We present the chapters in this manner to suggest that the authors, who might – on the surface – appear to have done very different inquiries into mathematics teacher education, could establish enough common ground to form AMTE Study Groups in order to develop collective inquiries at Borko's Phase 2 and/or Phase 3.

AMTE Monograph 5 Chapters

Authors of seven of the chapters in this monograph focus on the use of mathematical tasks as a launching point to help teachers better understand the content as well as the pedagogy related to that specific content. The differences between these chapters lie in the specific focus within the area of mathematical tasks as well as the teacher education model authors implemented around learning about and through mathematical tasks.

Goodman and Campbell tell the story of a statewide professional development academy for elementary teachers. Although the fundamental focus of the work within the academy stemmed from solving mathematics problems, collaborative investigations of those problems, and implementing those problems in their schools, the emphasis of this chapter is on the structure of the academy. In this chapter, Goodman and Campbell present a model for mathematics professional development that can be replicated at several sites around the country. An extension of the work they have presented in this chapter would be to conduct a study of teacher learning in several sites.

Goodman and Campbell could also seek to establish connections between their work and the work of Evans, Bean, and Romagnano. Similar to Goodman and Campbell, Evans and her colleagues report on a course in their rural schools program also revolves around specific mathematical tasks with the goal of increasing content knowledge while simultaneously focusing on pedagogy. The contexts of these two MTE programs differ, but the basic goals and methods appear to have more similarities than differences. Working together would cause the need for both sets of authors to examine and refine their underlying framework, potentially strengthening both models and creating a study at a higher Phase. Others who have an interest in distance education could team with Evans, Bean, and Romagnano to replicate their model for study at multiple sites.

Chval, Lannin, and Bowzer use some of the same theoretical constructs as Evans, Bean, and Romagnano, emphasizing the situated nature of the experience of implementing mathematical tasks as a launching point from which teachers learn mathematics and pedagogy. These authors suggest framing concepts that could influence others' choices of tasks to use with their MTE students, and in fact argue that we need this common framework in order to advance our MTE practices. Again, authors of these three chapters (and other interested MTEs) could utilize Chval et al.'s framework across several sites and study the impact on preservice teachers' learning.

An AMTE Study Group focused on the implementation of mathematical tasks might also consider working with the framework developed by Van Zoest and Stockero. These authors focus on the use of concentric task sequences, a structured approach to moving from a mathematical task to student thinking about that task and then to teacher thinking. Could other MTE's enactment of this model produce results similar to those of Van Zoest and Stockero?

Van Zoest and Stockero's notions of this structured approach bears some important similarities to the ideas underlying the problem solving cycle approach advocated by Koellner, Schneider, Roberts, Jacobs, and Borko. Here, however, video clips and student work stimulate the mathematical discussion that crosses over into student thinking and eventually into teacher thinking. Another model that encourages MTE students to travel through the cycle from mathematical task, through student work, and into teacher thinking is represented in the chapter by Hughes, Smith, Boston, and Hogel. This team of MTEs used written case stories and student work to launch the investigation. Here we have three sets of authors who are implementing similar trajectories to influence teacher knowledge. Could they join forces in an AMTE Study Group to design common data collection instruments that would allow an inquiry into learning at different sites with different, but somewhat similar, MTE models?

Clark presents a different model for engaging MTE students in learning about teaching – through learning about how to use the history of mathematics in secondary mathematics teaching. She provides enough detail for her model to be replicated at multiple sites. An AMTE Study Group on preservice teacher course design could support MTEs across the nation in designing and implementing the same course and then inquiring into what preservice teachers learn. This Study Group could also contain members who design and implement activities within a mathematics methods or content course, much as Hjalmarson and Suh did, and study their impact on teacher learning.

In more closely examining the MTE community's propensity for conducting Phase 1 inquiries, a few things are noteworthy. First, Phase 2 and Phase 3 inquiries require reaching across institutional boundaries. Some key barriers that will need to be circumvented include proximity, context, and theoretical constructs. Inquiries that have been conducted on teachers' collegiality and opportunities to collaborate indicate that lack of physical proximity as well as the structure of the work day (Rosenholtz, 1989; Taylor, 2004) constrain collegiality by limiting opportunities to meet face-to-face. The convenience and communication constraints found in this work on teacher collegiality apply to university/college faculty and other teacher educators as well. Proximity, however, can be purposefully bridged through the use of technology.

MTEs also cite the context of inquiries as negating the possibilities for collaboration. Contextual differences, such as course structure and program structure, can be overcome by focusing on what can be common across sites and reporting differences as the context of the study. These contextual differences add richness to a multi-site study by helping to sort out what aspects might be directly transferable to other contexts.

The theoretical and conceptual constructs, however, are another matter. Academic freedom and the need to publish original research both contribute to the scattered nature of the literature in mathematics teacher education. Academic freedom is important to innovation, but does not excuse the need to collaborate across institutional boundaries to enhance our MTE professional knowledge. Such inquiries would not only move the community towards adding reliable and generalizable knowledge to our profession, but also challenge inquirers to find common ground in terms of theoretical and conceptual constructs. Discussions that move inquirers to common ground are a place where the ideas of all researchers involved are refined, with the outcome of a purer and more useful lens. Such discussions could occur in AMTE Study Groups.

Overarching all of our work is the notion of MTEs' learning. In his chapter, Steele reports on his effort to model what it means to be a reflective practitioner, including open discussion of his pedagogical dilemmas and choices with those he was teaching. Cady, Hopkins, and Hodges focus on their own learning through a lesson study on one of the lessons that they implemented, one with preservice teachers and one with inservice teachers. Authors of both of these chapters could work together in an AMTE Study Group focused on MTEs' learning, collaborating to design inquiries across their contexts.

This monograph also contains chapters that contribute to our understanding of MTE in which the authors do not present information about inquiries into MTE. Instead these authors illustrate critical components of one aspect of MTE. Watanabe, Takahashi, and Yoshida carefully outline the nuances and critical understandings necessary to the study of instructional materials and the role of this process in one model of teacher development, Japanese lesson study. Niess, Ronau, Driskell, Kosheleva, Pugalee, and Weinhold discuss a much larger aspect of MTE: How do we prepare mathematics teachers to teach in an environment that is technologically saturated and constantly evolving in terms of technology? Both sets of authors provide detailed frameworks (or aspects of frameworks) that could be adopted by other inquirers and used to study mathematics teacher learning on a broader scale.

This monograph ends with a chapter about scholarship in mathematics teacher education, written by Reed and Mathews. In this chapter, the authors discuss the relationship of one's work in mathematics teacher education and the processes/standards for promotion and tenure, raising several issues with regard to our work as MTEs. Although different university/college contexts may employ very different standards and processes, many commonalities also exist. Building on the discussion started here by Reed and Mathews, an issue for future discussion in MTE, and a charge for a possible AMTE Study Group is: Are some university and college promotion and tenure committees more likely to reward inquiries that lie in different phases in Borko's framework? If so, what materials are necessary for educating department, college, and university colleagues of MTEs as to the value and constraints of inquiries at each phase of Borko's framework?

As editors of AMTE Monograph 5, we have enjoyed working with the chapter authors during the process of review, revision, and publication. We appreciate their patience and understanding as we worked to serve AMTE in this fashion. We are proud of this monograph and believe that it contributes to the effort of MTEs to share their work with each other. We hope that you find the contents helpful in reflecting on your practices as a mathematics teacher educator.

References

Ball, D. L. (1988). Unlearning to teach mathematics. *For the Learning of Mathematics, 8*(1), 40-48.

Borko, H. (2004). Professional development and teacher learning: Mapping the terrain. *Educational Researcher, 33*(8), 3-15.

Hiebert, J., Gallimore, R., & Stigler, J. W. (2002). A knowledge base for the teaching profession: What would it look like and how can we get one? *Educational Researcher, 31*(5), 3-15.

Lortie, D. (1975). *School teachers: A sociological study*. Chicago: University of Chicago Press.

National Mathematics Advisory Panel. (2008). *Foundations for success: The final report of the national mathematics advisory panel*. Downloaded 10/15/08 from http://www.ed.gov/about/bdscomm/list/mathpanel/index.html

National Research Council. (2001). *Adding it up: Helping children learn mathematics*. Washington, DC: National Academy Press.

Rosenholtz, S. J. (1989). *Teachers' workplace: The social organization of the schools*. White Plains, NY: Longman, Inc.

Sowder, J. T. (2007). The mathematical education and development of teachers. In F. K. Lester (Ed.), *Second handbook of research on mathematics teaching and learning* (pp. 157-223). Charlotte, NC: Information Age Publishing.

Taylor, P. M. (2004). Encouraging professional growth and mathematics reform via collegial interaction. In R. Rubenstein and G. Bright (Eds.),

Perspectives on the teaching of mathematics (pp. 219-238). Reston, VA: National Council of Teachers of Mathematics.
Taylor, P. M., & Ronau, R. (2006). Syllabus study: A structured look at mathematics methods courses. *AMTE Connections, 16*(1), 12-15.

Fran Arbaugh is an Associate Professor of Mathematics Education at the University of Missouri, where she teaches undergraduate and graduate preservice teachers as well as Master's and Doctoral students. She is widely published in the area of mathematics teacher education and will be joining the mathematics education faculty at Pennsylvania State University in August 2009.

P. Mark Taylor is Executive Director of Eagle's Nest Christian Academy, in Clinton, TN, where he also co-founded a publishing company, www.PublishWithKA.com, which publishes books on education and Christianity. He formerly served as an Associate Professor of Mathematics Education at the University of Tennessee.

Goodman, T. and Campbell, L.
AMTE Monograph 5
Inquiry into Mathematics Teacher Education
©2008, pp. 11-22

2

Linking Master Teachers and Mathematics Educators to Build a Statewide Professional Development Academy

Terry Goodman
University of Central Missouri

Larry Campbell
Missouri State University

In this chapter we present an overview of the Missouri Elementary Mathematics Leadership Academy, a cooperative effort designed to provide K-5 teachers with on-going, school-based professional development focused on enhancing teachers' mathematics understanding and use of related instructional strategies and resources. One important feature of the Academy was the collaboration between K-8 Master Teachers and mathematics educators. Working together, these two groups developed the overall Academy model, created and taught lessons designed to help teachers deepen their content knowledge, and facilitated teachers in developing professional development plans. We highlight the Master Teachers' contribution to the participants' growth as mathematics teachers.

In recent years, increased attention has been devoted to providing relevant, effective professional development for mathematics teachers at all levels. The demands and expectations of high-stakes state testing programs, the federal No Child Left Behind initiative, the rise of alternative certification programs, and school leaders' desire to implement standards-based mathematics curricula have highlighted further the need for professional development programs that enhance teachers' content and pedagogical knowledge and skills. Initiatives, such as the Mathematics and Science Partnership program, have funded a number of projects focused on providing such professional development.

In Missouri, one such project, the Missouri Elementary Mathematics Leadership Academy, has provided on-going, school-based professional development for K-5 teachers focused primarily on enhancing teachers' mathematics understanding and use of related instructional strategies and resources. In this chapter, we provide a brief overview of the project and then highlight what we believe to be the aspects of the project that have contributed most significantly to the participants' growth as mathematics teachers.

The Elementary Academy has been founded on the following three premises:

1. *Classroom mathematics teachers are the persons most responsible for creating an environment that is supportive of student learning and development.*

The *Professional Standards for Teaching Mathematics* (National Council of Teachers of Mathematics [NCTM], 1991) emphasizes two important points: (a) Teachers are key figures in changing the ways in which mathematics is taught and learned in schools; and (b) Changes in instruction require teachers to have thorough preparation, long-term support, and adequate resources. The Elementary Academy speaks to both of these points.

Furthermore, in *Principles and Standards for School Mathematics* [*PSSM*] (NCTM, 2000) the authors assert, "Students' understanding of mathematics, their ability to use it to solve problems, and their confidence in, and disposition toward mathematics are all shaped by the teaching they encounter in school" (p. 16-17). Helping elementary teachers deepen their mathematics understanding and develop more confidence in their ability to teach mathematics likely enables them to create a rich learning environment for their students.

2. *Elementary teachers' mathematics content knowledge has a significant effect on their ability to teach mathematics.*

Putnam, Heaton, Prawat, and Remillard (1992) studied how 5th grade teachers' mathematical knowledge affected their instruction and noted that "the limits of the teachers' knowledge of mathematics became apparent and their efforts fell short of providing students with powerful mathematical experiences" (p. 221). Sowder, Philipp, Armstrong, and Schappell (1998) found that teachers' practices changed as their content knowledge increased and deepened. Mewborn (2003) emphasized that "enhancing teachers' mathematical knowledge has been a component of a number of professional development projects, and the evidence overwhelmingly suggests that this is a crucial part of learning to teach differently" (p. 49).

3. *Teachers benefit most from professional development that is ongoing, school-based, and that includes mentoring/coaching from Master Teachers.*

Once again, the authors of *Principles and Standards for School Mathematics* (NCTM, 2000) state, "The work and time of teachers must be structured to allow and support professional development that will benefit them and their students" (p. 19). Stigler and Hiebert (1999) noted that "a requirement for beginning the change process is finding time during the workweek for teachers to collaborate" (p. 144).

Professional development needs to occur in a context that allows teachers to try, in their classrooms, what they have learned. Fennema, Carpenter, Franke, Levi, Jacobs, and Empson (1996) found that as teachers were able to validate the content and pedagogical knowledge they were learning through its use in their classrooms, they began to take ownership of the models they were studying. Schifter (1998) found that it was crucial for a professional development project to span a school year so that teachers could try out, in their own classrooms, what they were learning.

Similarly, Stein, Silver, and Smith (1998) found that teachers need safe, supportive environments in which to discuss issues of content and pedagogy with peers. Models for continuous improvement of teaching used in Japan and China feature "long-term, school-based reform in a community of learners with opportunities to grapple with significant mathematical ideas and to consider how students engage with these mathematical ideas" (Mewborn, 2003, p. 50).

There are three main objectives, then, of the Missouri Elementary Mathematics Leadership Academy:

1. To create regional elementary mathematics academies that focus on enhancing the mathematical content knowledge of K-5 teachers.
2. To create opportunities for K-8 Master Teachers, working with higher education mathematics and mathematics education faculty, to be involved in providing K-5 teachers with on-going, school-based professional development. This professional development focuses on enhancing teachers' ability to teach key elementary mathematics concepts and strands, use related instructional strategies, and use a variety of instructional resources and technologies.
3. To provide inservice teachers with classroom-based mentoring and coaching from Master Teachers.

Overview of Project Activities

Since 2005, the Missouri Elementary Mathematics Leadership Academy has provided both two-week summer content academies and a one-week summer pedagogical academy for K-5 teachers (86 teachers at one site in the first year, 120 teachers at two sites in the second year, and 165 teachers at three sites in the third year). Teachers from the same school or district attended as teams selected by the local districts. The content workshops focused on the number/operations and geometry/measurement strands of the elementary mathematics curriculum. Each workshop consisted of eight lessons developed and taught by Master Teachers who were recruited and trained by the project Co-PIs (university mathematics education faculty). All of the Master Teachers (with the exception of two university faculty members) were current or former elementary and/or middle grades teachers.

The summer content academy lessons were built around mathematical investigations in which the elementary teacher participants explored the "big ideas" of number and operations in one year or geometry and measurement in the other. Because these lessons focused on helping teachers deepen their mathematical understanding, the participants were often "stretched" to think about ideas in new ways, to look for relationships and generalizations, and to consider the connections among various concepts.

Generally, four to five Master Teachers taught the lessons to a team of 20 elementary participants. The Master Teachers also developed "team builder" activities and assisted the participants with their homework. All participants at an Academy site met together in two large group sessions (first session each morning and last session each afternoon) where the project Co-PIs led them in activities and discussions that focused on problem solving, use of a variety of instructional strategies and resources, and other day-to-day issues arising from a given lesson.

Later each summer, all Academy participants met at one site for a one-week pedagogical academy that had two foci. First, each team of elementary teachers developed, for the next academic year, professional development goals related to what they had learned in the earlier content workshops. Each team also planned specific activities for the year that would enable them to reach their own professional development goals. Teams were assisted in their planning by one of two project Regional Teacher Enhancement Coordinators (RTECs), former classroom teachers who work with each participant team throughout the school year to help them implement their professional development goals and activities.

The second focus of the pedagogical academy was a workshop lead by Doug and Barbara Clarke, mathematics educators from Australia. Clarke & Clarke (2004) helped develop the *Early Numeracy Interview Booklet* (Department of Education, Employment and Training, Victoria, 2001) that contains an instrument elementary teachers can use to determine an individual student's understanding of fundamental "growth points" relative to number/operations and geometry/measurement. Academy participants received information and training regarding the development, use, and interpretation of results from the Inventory. Teams also had the opportunity, during this third week, to use the Inventory with at least one child and Growth Points kits were provided for teams to take back to their schools.

During the academic year following a given summer Academy, teams were encouraged to use ideas, strategies, and resources from the summer experience with their students. Each team also continued to engage in activities to help meet their professional development goals, including lesson study, curriculum development and implementation, and alignment of local curricula to state and national standards. The RTECs met with each team on a monthly basis and provided feedback and mentoring for the teachers as they attempted to implement new activities, instructional strategies, and resources in their classrooms. The project provided funding for substitute teachers so that teams could meet for planning one-half day each month.

The diagrams found in Figure 1 and Figure 2 delineate important components of the project. Figure 1 illustrates the Academy staff organization and roles; Figure 2 illustrates the organization of project activities.

Missouri Elementary Mathematics Academy Staff Organization and Roles

Co-PIs

1. Select and train project Master Teachers.
2. Coordinate planning and implementation of all project activities.
3. Work with RTECs to plan and coordinate academic year activities.
4. Work with Evaluator to plan and coordinate project evaluation.

↕ ↕ ↕

Regional Teacher Enhancement Coordinators (RTECs)	Master Teachers (MTs)	Project Evaluator
1. Provide orientation to the Academy for participant teams 2. Assist participant teams with planning of academic year professional development goals and activities 3. Meet with participant teams during the academic year to facilitate implementation of professional development goals and activities	1. Develop content focus, activities, and resources for summer Academy lessons 2. Teach summer Academy content lessons to a group of 20 elementary participants 3. Some MTs serve as mentors for participant teams from their districts	1. Develop pre- and post-tests to measure participants' content knowledge 2. Develop pre- and post-surveys to assess participants' dispositions toward student learning, instructional strategies, use of instructional resources, and assessment

↕ ↕

Participant Teams

1. Participate in a two-week summer content Academy for each of two consecutive summers
2. Participate in a one-week summer Growth Points workshop for each of two consecutive summers
3. Develop, plan, and implement professional development goals and activities for each of two academic years
4. Complete all project evaluation instruments and, where applicable, report results of their use of the Growth Points Inventory

Figure 1. – Staff Organization and Roles.

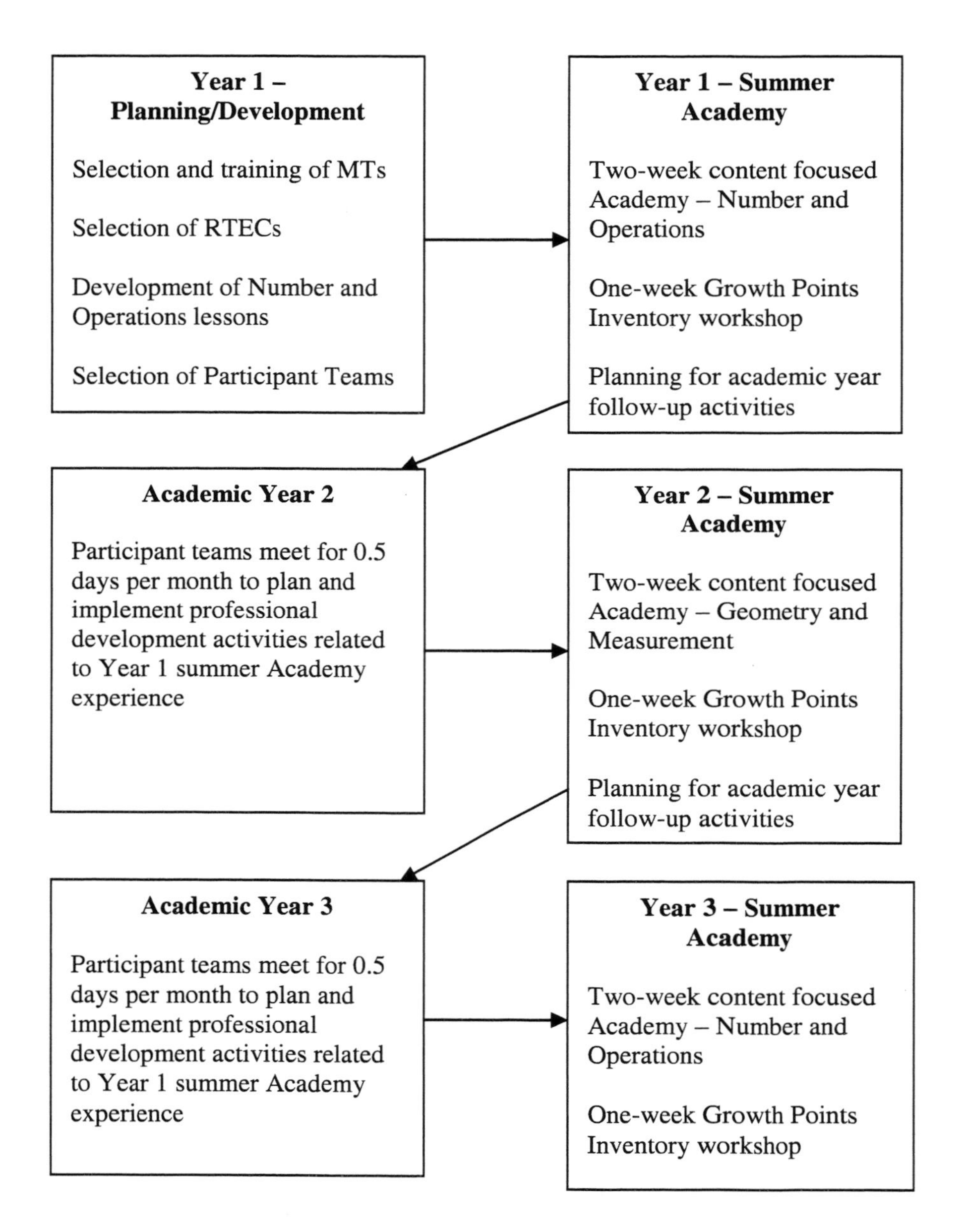

Figure 2. Organization of Activities.

Impact on Project Participants and Master Teachers

Project participants in both summer academies demonstrated significant gains in their mathematics content understanding. For Year 1, pre- and post-tests created by the project staff were administered to the participants at the beginning and end of the two-week content Academy. For Year 2, the *Diagnostic Teacher Assessment in Mathematics and Science for Geometry and Measurement* (University of Louisville Center for Research in Mathematics and Science Teacher Development, 2004) was used as a pre- and post-test. In Year 3, the evaluator developed a test assessing procedural knowledge of number and operations and also administered the 2004 *Elementary Test for Number and Operations* from the Learning Mathematics for Teaching (LMT) Project at the University of Michigan. In all three years, statistical analyses of total scores for teacher participants indicated significant gains in content knowledge for number/operations and geometry/measurement (both procedurally and conceptually). Further, on a post-Academy questionnaire given at the end of the second and third summer Academies, between 91% to 96% of the teachers indicated that the lessons were "useful" or "very useful" relative to their professional development goals, would impact "to a great extent" their own understanding of geometry and measurement and their confidence in teaching mathematics, and would improve their ability to use mathematical language accurately and carefully.

Project participants strongly indicated that the Academy experiences would enhance their instruction and ultimately their students' understanding of mathematics. The participants received training in the use of the Growth Points Inventory and we encouraged them to use this instrument with some of their students. Several teachers used this instrument with students during year 2 of the project; the majority of the participants used the Inventory with students during year 3. Teachers who used the Inventory indicated that it provided them with an in-depth assessment of a student's mathematical understanding; the majority of students who took a pre- and post-inventory in year 2 of the project showed significant growth in numeracy. As one teacher observed, "Tasks that were well below the student's grade level could not be done by the student. Sometimes we assume they know it and never address it in class." Another teacher reported, "Interviewing a student gives the teacher a lot more information on achievement levels than a paper/pencil test." A third teacher said, "It [Growth Points Inventory] is a great tool for helping me identify where lower ability students are performing." A fourth teacher reported that what a student knew was "different" from what the teacher expected and said, "The student applied different ideas than we'd practiced in the classroom."

Ultimately, as teachers investigated what students did and did not understand about the big ideas of number/operations and geometry/measurement, they broadened and deepened their own understanding of those same concepts to "make sense" of what their students were doing. This new understanding then prompted the teachers to consider how they could modify their teaching to address student understanding and misunderstanding of important mathematics concepts and relationships in better ways than they had typically done. The teachers were most enthusiastic in their evaluation of Drs. Clarke and the Growth Points Inventory training. One teacher responded by saying, "The activities and stations during the time with Doug and Barbara were so engaging and will be helpful to take back to our school. It let us think outside the box!"

Participants also engaged in a variety of explorations that modeled the kinds of student investigations, instructional strategies, and instructional resources (a variety of manipulative materials, Geometer's Sketchpad, etc.) that would help them create dynamic, student-centered learning environments. On a post-Academy questionnaire from Year 2, teacher participants strongly indicated that their Academy experiences would enhance their ability to teach mathematics well; enhance their ability to choose or create good examples, problems, and assessments for geometry and measurement; enhance their ability to use technology and Geometer's Sketchpad; and enhance their ability to use pedagogical methods such as collaborative or inquiry-based learning.

One project goal has been to develop a model that would allow K-8 Master Teachers to become directly involved in providing professional development for both preservice and inservice teachers. Although we expected that participation in the Academy would provide professional development for the Master Teachers (MTs), we were surprised by the strength of that result. The MTs indicated that writing and teaching the Academy lessons required them to think deeply about the mathematical concepts and relationships found in those lessons. Creating a learning environment for the elementary teachers that would "stretch" them and, at the same time, be "safe" was a challenge for the MTs. Many of the MTs had never taught other teachers and this Academy experience was a profound learning experience for them. Nearly all the MTs indicated that, as a result of their participation in the Academy, they now had more confidence in their ability to provide leadership and professional development for their colleagues and in their schools and districts.

Implications for Professional Development

Based on our Academy experiences, observations, and evaluations, we can identify several implications for professional development of inservice mathematics teachers. We hope that readers find these implications applicable to their own work with such individuals.

The Master Teachers' contributions to the quality and depth of the elementary teachers' experiences have been vital to the project. MTs' work in writing and teaching the content lessons helped the elementary teacher participants explore mathematics concepts in a safe and supportive environment. Although the elementary teacher participants viewed the lessons as challenging for them, they also felt comfortable because the MTs provided consistent assistance and encouragement. The teachers' "comfort level" is evidenced by the fact that 80 of the 86 Year-1 participants returned for Year 2 of the Academy. The six teachers who were not able to return expressed their desire to continue, but personal factors did not allow them to do so. For Year 3, the number of participants grew to over 160, with a significant waiting list for the Academy. We believe the teachers' enthusiasm for participating in the Academy has been significantly influenced by the work of the MTs. Logistically, it might have been easier for the lessons to be written and taught by higher education faculty, but we believe the MTs added "credibility" to the Academy lessons and activities. The impact of the MTs' teaching, coaching, and mentoring was consistent with earlier research results from Stein, Silver, and Smith (1998) and Mewborn (2003). Teachers teaching other teachers can be a powerful professional development model.

The Growth Points Inventory provides elementary teachers with an instrument/tool that can be used to determine individual student's depth of mathematical understanding. This tool supported teachers in considering how their students think about mathematics and helped move the teachers away from simply looking at what facts students can memorize and/or what procedures they can imitate. Further, in using this tool, the teachers had to think deeply about their own mathematics understanding, not just the mathematical procedures their students have mastered. Indeed, the Growth Points Inventory provided project teachers with a wonderful follow-up to the content lessons they explored and helped them "apply" what they learned through participating in the project to their work with their students. Effective professional development for K-5 teachers should link growth in teachers' content understanding to their students' mathematical thinking.

Clarke and Clarke (2004) found that effective teachers:

- Structure purposeful tasks that enable different possibilities, strategies, and products to emerge;
- Focus on important mathematical ideas and make the mathematical focus clear to children;
- Choose tasks that engage children and maintain their involvement; Encourage children to explain their thinking and build on children's ideas and strategies; and
- Use teachable moments as they occur.

Working with the Growth Points Inventory helped the Academy participants begin to develop the skills and insights suggested by Clarke and Clarke.

Helping teachers explore deeply the mathematics of number/operations and geometry/measurement provides a foundation for further discussions and explorations of related topics, such as alternative assessment strategies, effective instructional strategies, and the use of concrete materials and technology in the elementary classroom. As the teachers' understanding of the mathematics deepened, their formal and informal discussions seemed to focus naturally on students' thinking. In addition, they began to focus on classroom implications of what they were learning in the Academy. A study by Gearhart, Saxe, Seltzer, Schlackman, Ching, Nasir, et al. (1999) indicated that children learned more mathematics in classrooms where instruction (a) was based on students' ways of thinking, (b) engaged students in problem solving with rich problems, and (c) assisted students in seeing the underlying links among various mathematical concepts and symbols. In the light of these findings, the Academy lessons often focused on student thinking, problem solving, and mathematical connections. As PIs of the project, we believe effective professional development for K-5 teachers should focus on providing a learning environment for them that is consistent with the classroom environment we would encourage them to create in their own classrooms.

The follow-up activities and on-going support provided for the teachers during the academic year contributed significantly to their professional development. Recommendations from the National Council of Teachers of Mathematics (1991) indicate that professional development projects have a responsibility in "supporting teachers in self-evaluation and in analyzing, evaluating, and improving their teaching with colleagues and supervisors" (p. 181). This on-going activity and support helped the teachers maintain the "momentum" generated by their summer Academy experiences. The RTECs provided the teachers with consistent and relevant feedback, encouragement, and mentoring throughout the year, so that the teachers were able to continue their professional growth and did not have to wait for the next summer Academy to get "recharged." Indeed, many of the teachers have continued to work together as teams within their schools or districts even after their participation in the Academy ended. Many of the teachers now have the skills, knowledge, and confidence needed to continue to monitor and be responsible for their own professional development. We believe effective professional development must be more than something that happens in a summer experience; school-based, on-going activities and support are vital.

References

Clarke, B., & Clarke, D. (2004). Mathematics teaching in grades K-2: Painting a picture of challenging, supportive, and effective classrooms. In R. Rubenstein & G. Bright (Eds.), *Perspectives on the teaching of mathematics.* (pp. 67-81). Reston, VA: National Council of Teachers of Mathematics.

Fennema, E., Carpenter, T. P., Franke, M., Levi, L., Jacobs, V., & Empson, S. (1996). A longitudinal study of learning to use children's thinking in mathematics instruction. *Journal for Research in Mathematics Education, 27*, 403-434.

Gearhart, M., Saxe, G. B., Seltzer, M., Schlackman, J., Ching, C., Nasir, N., et al. (1999). Opportunities to learn fractions in elementary mathematics classrooms. *Journal for Research in Mathematics Education, 30,* 286-315.

National Council of Teachers of Mathematics. (1991). *Professional standards for teaching mathematics.* Reston, VA: Author.

National Council of Teachers of Mathematics. (2000). *Principles and standards for school mathematics.* Reston, VA: Author.

Mewborn, D. S. (2003). Teaching, teachers' knowledge, and their professional development. In J. Kilpatrick, W. G. Martin, & D. Schifter (Eds.), *A research companion to Principles and Standards for School Mathematics* (pp. 45-52). Reston, VA: National Council of Teachers of Mathematics.

Putnam, R. T., Heaton, R. M., Prawat, R. S., & Remillard, J. (1992). Teaching mathematics for understanding: discussing case studies of four fifth-grade teachers. *Elementary School Journal, 93,* 213-228.

Schifter, D. (1998). Learning mathematics for teaching: from a teachers' seminar to the classroom. *Journal of Mathematics Teacher Education, 1,* 55-87.

Sowder, J. T., Philipp, R. A., Armstrong, B. E., & Schappelle, B. P. (1998). *Middle grade teachers' mathematical knowledge and its relationship to instruction: A research monograph.* Albany, NY: State University Press.

Stein, M. K., Silver, E. A., & Smith, M. S. (1998). Mathematics reform and teacher development from the community of practice perspective. In J. Greeno & S. Goldman (Eds.), *Thinking practices: A symposium on mathematics and science learning* (pp. 17-52). Hillsdale, NJ: Erlbaum.

Stigler, J., & Hiebert, J. (1999). *The teaching gap: Best ideas from the world's teachers for improving education in the classroom.* New York: Free Press.

University of Louisville. (2004). *Diagnostic Teacher Assessment in Mathematics and Science for Geometry and Measurement.* University of Louisville Center for Research in Mathematics and Science Teacher Development. Louisville, KY.

University of Michigan. (2004). *Elementary Test for Number and Operations.* Learning Mathematics for Teaching Project. Ann Arbor, MI.

The professional development project described in this chapter, titled "Missouri Elementary Mathematics Leadership Academy," was funded by the U. S. Department of Education and the Missouri Department of Elementary and Secondary Education through the *Mathematics and Science Partnership* program.

Terry Goodman is Professor of Mathematics Education at the University of Central Missouri where he teaches courses for pre- and inservice, K-12 mathematics teachers. He serves as Co-PI for the Missouri Middle Grade and Secondary Mathematics Leadership Academies. His primary areas of research interest are problem solving and professional development.

Larry Campbell is Professor of Mathematics at Missouri State University where he enjoys working with students and future teachers with mental blocks in mathematics. He serves as Co-PI for the Missouri Middle Grade and Secondary Mathematics Leadership Academies. He is a frequent speaker and workshop leader at professional conferences.

Evans, B., Bean, H., and Romagnano, L.
AMTE Monograph 5
Inquiry into Mathematics Teacher Education
©2008, pp. 23-33

3

Toward a Situative Perspective on Online Learning: Metro's Mathematics for Rural Schools Program[i]

Brooke Evans
The Metropolitan State College of Denver

Hamilton Bean
University of Colorado at Boulder

Lew Romagnano
The Metropolitan State College of Denver

The No Child Left Behind Act [NCLB] (U. S. Congress, 2001) requires "highly-qualified teachers" in every classroom. This requirement is difficult for rural schools to fulfill due to their geographic isolation, small student populations, limited teaching staffs, and the need for teachers to teach a variety of subjects. In this chapter, we describe an online, inservice mathematics teacher education program for rural teachers guided by a situative theory of learning. Preliminary research on the program suggests that an online learning environment can be used to create connections among rural teachers and facilitate collaborative, problem-based, process-centered instruction.

Meeting the "highly-qualified teachers" requirement of the No Child Left Behind Act [NCLB] (U. S. Congress, 2001) is difficult for rural schools to achieve due to their geographic isolation, small student populations, limited teaching staffs, and the need for teachers to teach a variety of subjects. As a result, interactive distance learning has been proposed as a way to help rural teachers meet NCLB requirements (Reeves, 2003). In 2005, the Mathematics Teaching and Learning Group at the Metropolitan State College of Denver (Metro) began development of Metro's Mathematics for Rural Schools Program (Metro's Program), a three-course series designed to enhance the content knowledge of rural K-12 mathematics teachers in Colorado. Metro's Program uses web-based technology to make these content-oriented courses available to teachers. Although the courses are content driven, the delivery of the courses exemplifies teaching practices that are directly transferable to K-12 mathematics classrooms, supports the development of mathematical proficiency, and fosters a mathematical community among teacher-learners. The courses are available for graduate credit for practicing teachers and are intended to serve as a model for content-centered professional development for K-12 mathematics teachers in rural schools across Colorado. In this chapter, we elaborate how Metro's Program draws on a situative theory of learning to foster collaborative, problem-based, and process-centered instruction.

Guiding Principles

Mathematics knowledge for teaching is specialized knowledge of the big mathematical ideas that underlie the K-12 curriculum and what it means to do mathematics. Solving novel problems, explaining or justifying a solution or strategy, extending or generalizing a result, and unpacking complex mathematical ideas, strategies, and procedures are the core elements of this approach (Ball & Bass, 2003). When designing the courses, we incorporate these elements within a situative framework. A situative view of learning asserts that knowledge is created through social interactions as learners participate in a community of practice (Greeno, 2003). The relationship between individual learners and the community is reflexive; what individuals learn is inseparable from the context in which it is learned, and the learners influence the context through their negotiation of meaning (Peressini, Borko, Romagnano, Knuth, & Willis, 2004). In our situative approach, we emphasize the learning of mathematics through *participation* within a community of teacher-learners. In other words, a primary goal of Metro's Program is to have participants "contribute to mutual understanding by appreciating and explaining assumptions involved in their thinking and that of other participants" (Greeno, 2003, p. 316).

With this principle in mind, Metro faculty attempt to model, online, an approach to teaching that is successfully used in the faculty's campus-based courses. To accomplish this objective, Metro's Program has to retain two essential components of the campus-based courses: 1) a problem-driven curriculum, and 2) a collaborative instructional approach that emphasizes reasoning, communication, and creation of a learning community. The challenge is to maintain the integrity of these course components while moving to an online environment that reaches Colorado's rural teachers.

We argue that an essential element of learning mathematics for teaching is engaging teachers in mathematics problem solving in a manner similar to that which they are expected to use with their own students. Teachers and students alike need the experience of struggling with mathematics problems, using their instincts, decomposing mathematical ideas, understanding the "whys," and making and testing conjectures (Mewborn, 2003). In addition, we agree with Ball and Cohen (1999a), who assert that "[c]reating and sustaining an inquiry-oriented stance...is a social enterprise" (p. 17); therefore, small group interaction and collaboration among teachers are key facets of Metro's Program. According to Stigler and Hiebert (1999), "Increased collaboration allows teachers to develop a 'shared language' for describing and analyzing mathematics and the teaching of mathematics" (p. 123). Metro's Program attempts to facilitate a collaborative environment for doing mathematics through the use of web-conferencing software that allows participants to interact in a variety of ways in real time. [ii]

Course Description

As of May 2007, three courses had been offered to teachers in the San Luis Valley of Colorado. The third course—the focus of this article—was conducted in a similar manner to the first two. The course consisted of two on-line meetings per week for five weeks, situated between day-long, face-to-face opening and closing sessions. For the on-line sessions—facilitated by Metro faculty located in Denver—the teachers worked in groups of two to four people, either from schools or homes spread throughout the San Luis Valley. Some groups were in the same room sharing one computer; other groups were "virtual," i.e., individuals were in separate locations, each using his or her own computer. Roughly forty percent of the participants in the third course had participated in either one or both of the previous courses and were therefore familiar with the course procedures, the web conferencing software, the Metro faculty, and each other. A handful of the participants were also familiar with *Geometer's Sketchpad*®—a software tool used during the third course.

The third course began with a full-day, in-person kick-off session held on a Saturday in a hotel conference room in Alamosa, CO, some 230 miles from the Metro faculty members' home base in Denver. A few of the 20 course participants traveled more than 70 miles to attend. The initial in-person session was important to establish the course community and the interactional norms among the participants. During the session, Metro faculty introduced themselves, outlined the course goals, and familiarized the participants with the course's content, pedagogical elements, and technology.

At one point during the session, a Metro faculty member—Don—stepped into the middle of the conference room and held up a tetrahedron constructed from six toothpicks and four small marshmallows. Don instructed the participants—who were seated in groups around folding tables—to make their own tetrahedra from the toothpicks and marshmallows provided. Don said, "Find the surface area and volume of this thing" as though it were an inconsequential problem. The participants rapidly fell into their expected roles for the course and worked collaboratively to unpack the mathematics of the problem. Participants quickly realized that the *process* of finding an answer and discovering the underlying mathematical ideas was more important than the answer itself. Nearly three hours later, most of the groups were still working on the problem and getting ready to present their insights and solutions. During presentations, participants were asked to restate and/or analyze another group's solution process.

This in-person session established the form and "feeling" for the online sessions of the course. Camaraderie, collaboration, and (occasionally) friendly competition were brought into the online environment through the use of web conferencing software. The online sessions occurred from 5:00 pm to 6:45 pm twice per week. The four Metro faculty used their laptop computers and microphone headsets and split into two pairs; one pair facilitated a session for elementary teachers (12 teachers), and the other pair facilitated a session for secondary teachers (8 teachers). The exact problems and proposed learning outcomes were slightly different for each group based on their needs, but both sessions incorporated the same basic content. Online sessions were often conducted from one of the Metro faculty members' homes in order to accommodate care of young children (this situation reinforced community ties with several course participants who also logged on from home in order to care for their children).

Once everyone was online, Metro faculty used the software tools to present interesting, complex problems for the groups to tackle. Metro faculty then engaged in electronic "eavesdropping" in order to listen to group interactions, observe "whiteboard" work, and provide hints and encouragement. Metro faculty subsequently brought the groups together (virtually) to restate and analyze solutions and to engage the underlying mathematical concepts. Problems for the courses were chosen to expand the teachers' mathematical thinking concerning the K-12 curricula and to communicate their mathematical ideas. Groups generally used synchronous voice, text chat, and the "whiteboard" to communicate with each other. Figure 1 contains a "screenshot" from a typical problem-solving session.

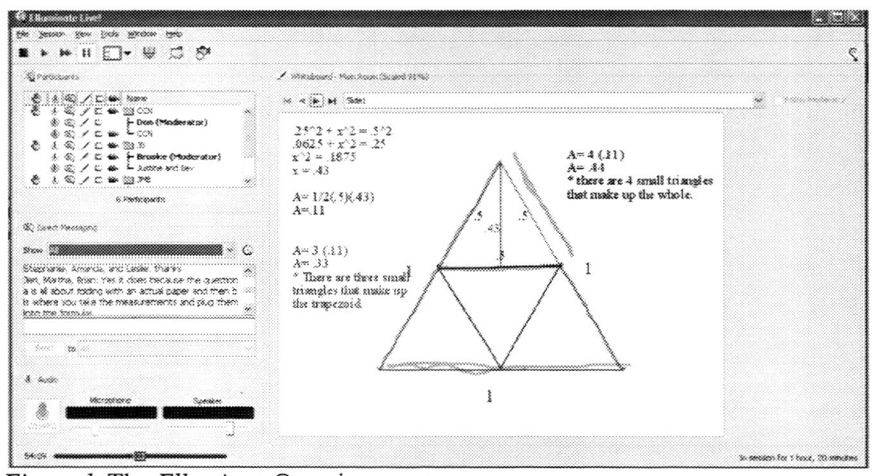

Figure 1. The *Elluminate®* environment.

The objective of the course was to provide mathematics content for teachers while simultaneously modeling a pedagogical approach designed to create a community of teacher-learners. During the "debrief" at the in-person closing session, participants noted how the course had fostered collaboration and communication within the San Luis Valley mathematics teacher community. Working with colleagues can help teachers answer questions, learn new ideas, and reflect on their teaching practices (Romagnano, 1994). The course participants also formed bonds with Metro faculty, furthering communication, collaboration, and reflection. As Sowder (2005) notes, direct communication with a mathematics education specialist is important in pushing teachers' mathematical thinking.

Research Question and Methods

Although Metro faculty hoped that the course would serve as a model for how to implement situative pedagogical techniques in K-12 classrooms, pedagogy was not *directly* emphasized in the course. Metro faculty expected that teachers would, ideally, perceive the value of building community and seek to transfer such knowledge to their classrooms. Also, teachers are often *told* that a problem-solving approach creates meaningful experiences for students, but many teachers have not had experience with the approach as learners themselves. As previously noted, we argue that an essential component of learning mathematics for teaching is to engage teachers in mathematics problem solving in ways that they can use with their own students. Experiencing a new approach firsthand is critical if teachers are to see its value and become comfortable using it. Based on this premise, we developed a research question, namely: *Are teachers building community and changing their teaching practices as a result of Metro's Program?* A preliminary assessment of the program, using a combination of observations, questionnaires, and focus group interviews, suggests that Metro's Program has substantively influenced teachers' thinking about their work in their classrooms.

Preliminary Findings

Participants in the third course completed pre- and post-course questionnaires. Pre-course questionnaires focused on teachers' reasons for enrolling in the course and what they hoped to gain from it. Post-course questionnaires were used to assess whether, as a result of the course, teachers: 1) felt more comfortable teaching mathematics; 2) spent more time preparing mathematics lessons; 3) perceived that they engaged more successfully with students around mathematics issues; 4) believed that they were able to prepare mathematics lessons more effectively; and 5) had attitudes toward teaching mathematics that had become more positive. Mean values (N = 19) for these elements are provided in Table 1. These preliminary data from self-reports suggest to us that teachers found the course beneficial.

Table 1: *Post-Course Assessment*

Assessment Category	Mean Score (1-7 scale)
I am more comfortable teaching mathematics as a result of this course.	5.6
I devote more time to teaching math as a result of this course.	5.2
I find ways to engage my students more successfully as a result of this course.	5.6
I prepare my math lessons more effectively as a result of this course.	5.4
I have a more positive attitude toward teaching math.	6.0

Note: 1 is low and 7 is high

In addition to questionnaires, focus group interviews provided data indicating that the course encouraged teachers to revisit basic assumptions about mathematics instruction and spurred adjustments to teaching practices in three interrelated ways: 1) teachers are attempting to utilize the problem-based approach modeled during the course; 2) teachers are emphasizing the importance of the learning process itself, rather than simply disseminating content to students; and 3) teachers are networking with course participants to strengthen community bonds and to share ideas and resources.

Teachers' Use of the Problem-Centered Approach

Teachers found the problem-centered approach stimulating, and several indicated that they would use it in their own classes. One teacher commented, "What is most useful is how well [the problem-centered approach] transfers to my classroom." Key to the success of the approach is the "fun" teachers had in trying to solve the problems. "[Problems] were stimulating and exciting to find the answers to. I just felt really great when I could solve it, or even when someone else solved it. It was like, 'Wow, look at that!'" Another teacher stated, "There was one problem I even lost sleep over trying to figure it out....When a problem gets into you like that and really makes you think, that's really positive. I like a challenge." Teachers recognized the value of the problem-centered approach to motivate and engage their students. One teacher commented, "I enjoy learning better ways to teach from watching you all teach us. I even try to use your techniques in my chemistry class."

Teachers also recognized the value of asking students to rephrase their own and others' approaches to solve a given problem. "Some problems weren't really challenging until [Metro faculty] made us rephrase them...They'd put a twist on it and then it became engaging again." Small group interaction enhanced the problem-centered approach. One teacher stated, "I loved looking at how other people solved problems." Another added, "I liked being able to see how others approached the problems. That really helped me grow as a teacher. I feel I have a wider range of approaches to teach." Comments such as these affirm the value of collaboration.

Teachers' Focus on Process. The course underscored the importance of the process of learning for many teachers. As one teacher observed, "I liked how the moderators guided us in our learning instead of just telling us the answer." Another teacher added, "It wasn't getting the answer that was important; it was the process of doing it." This teacher began experimenting with the process-centered approach in his classroom by "allowing children to explain the process by which they're doing things and allowing them to see that there are different ways of getting the same answer." The process-centered approach appears to have also affected change in other participating teachers' classrooms. As one teacher stated, "A light would go on [for someone] when the rest of us were stumbling in the dark. It made me aware of how important it is to reach children on lots of different levels, to reach all the children." Another added, "I find it's more important to teach the 'why.' They [Metro faculty] kept saying, 'We want to know the why.'" Another teacher summed up the value of the process-centered approach when she stated, "If you can teach [students] that inquiry, they're going to be much better learners overall....Experience like that is like a branding iron to the brain."

Although the teachers valued the process-centered approach, interviews revealed teachers' concerns about "helping" students versus "casting them adrift." Teachers generally praised the course for allowing them "to do [their] own thinking" and letting them "explore and find the answers;" yet others stated, "I have students who'll say, 'Just give me the formula'" or "Could my students stand going through the process, when in actuality, half of them are just going to want the formula?" In other words, teachers stressed the importance of the process-centered approach for their own learning, but some were reluctant to try such an approach in their classrooms. It is not unreasonable for teachers to want to package mathematics for their students in a way that will increase the probability that students will get the "right answers," but according to Ball and Cohen (1999b), the risk is that "what [teachers] have accomplished is perhaps less major in terms of learning" (p. 11). Metro's Program creates this tension, or at least increases it, because it puts teachers in the position of having to reconcile what they see as a valuable community of practice for their own learning while resisting creating such a community for their students. However, several of the teachers have, as a result of the course, expressed more willingness to move toward process-centered approaches.

Influence on Community

Participants acknowledged the benefits of working with other teachers in their community. One teacher stated,

> The camaraderie is one whole piece [of the course's success]. Being able to work with my colleagues who are in some other room all day and I never see, and to learn how their minds work, is just wonderful. We worked with three different schools.

The course format allowed teachers to sit "elbow-to-elbow." One teacher commented that "it was really juicy to sit at the same table with other people." Another teacher added, "There's a big learning process in the interaction....You get a lot more out of it if you can interact more." One teacher commented that Metro's Program brought rival schools closer together. These teachers even agreed to travel to one location in order to work together as a single online group. Another teacher acknowledged the benefits of collaboration, but stated, "When you throw it into the real schedule—in reality, we're all on our own islands." Another teacher, however, pointed out that as a result of Metro's Program "you increase your comfort level with being able to contact somebody [in a different school]." As mentioned previously, the teachers saw benefit from sharing ideas and taking on the thinking of others. One teacher mentioned that as a result of the course, their school administration has recognized the importance of teacher communication and collaboration and is looking for new ways to foster both.

Conclusion

The preliminary data suggest that Metro's Program has created, in an on-line environment, a community of practice in which teachers and their Metro faculty instructors learned to participate in constructive mathematics activities. This participation took at least two forms: interaction within small groups (both real and virtual) to solve problems, and interaction across groups and with Metro faculty to explain and justify mathematical work and compare or critique the work of others. The teachers' mathematics learning manifested itself as these patterns of participation developed over time and, as Peressini et al. (2004) argue, this situative learning directly influences "teachers' developing knowledge and beliefs about mathematics, mathematics-specific pedagogy and professional identity" (p. 73).

Mathematics learning environments designed with a situative view of learning can be difficult to put into practice because a situative view challenges traditional assumptions about the teacher's role in the classroom and requires students (and teachers) to struggle in unfamiliar ways. This approach also takes time—a rare resource for teachers. One teacher commented that "students would have fun playing [with *Geometer's Sketchpad*®] and figuring it out...if we had the time." Another teacher added, "It does come down to time. I certainly would not want to present [*Geometer's Sketchpad*®] with my limited experience with it." Time is not only needed for a problem-based, process-centered approach, but is also needed in order to build community. A familiar dilemma highlighted in our work is that teachers fully recognize the considerable benefits of community but feel they cannot adequately work to build community in their own classrooms because of a lack of time.

Several participants also indicated "how important it is to have other [teachers] in the classroom to reach students in different ways, and to hear." Such comments may reveal an implicit reluctance to depend on *students* for multiple perspectives and feedback. One of the goals of Metro's Program is to reduce teachers' anxiety about not presupposing the "right" perspective; we support this goal by allowing them to use their own instincts to make and test conjectures (Mewborn, 2003). Yet such a position requires a high tolerance for ambiguity as an instructor. One teacher acknowledged that some of her colleagues would be opposed to the teaching philosophies underlying Metro's Program. Despite these challenges, Metro faculty are encouraged that participants are reflective about issues of control and process within their classrooms. This dynamic has influenced the Metro faculty's own instruction in that faculty members are more explicit in modeling the types of interaction they would like to see from participants.

Finally, it is important to mention the role of the technology. Participants praised the technology for facilitating interpersonal relations—especially when the technology was transparent—and criticized the technology when it degraded contextual cues or became a barrier to communication. Teachers felt that technological hurdles, including variable online connection speed, microphone feedback and distortion, absence of certain mathematical symbols in the software tools, and software functionality issues, occasionally hampered interaction. As one teacher stated, "I could spend my time focusing on the whiteboard and getting it to work right, or focus on the math."

Next Steps

From our perspective, a significant challenge facing Metro's Program is the emphasis teachers have placed on the pedagogical elements of the courses. Although teachers' interest in pedagogical elements was not unexpected, the *intensity* of that interest warrants reconsideration of the Program's exclusive focus on content. Metro faculty strongly believe that maintaining a focus on mathematics content is essential, and that references to pedagogy, if any, should remain secondary. However, the teachers' comments indicated that the course's tightly facilitated online sessions perhaps obscured that the courses are based on an extensive set of problem-centered lessons that have been refined through years of combined teaching. As one participant stated, "[Metro faculty] knew just what type of hint to give you or what kind of question to ask to make you go a little further." Our concern is that teachers who have completed the course risk deploying their own problem-based, process-centered approaches without adequate understanding of the background and preparation underlying the courses. To address this issue, Metro faculty are periodically planning to incorporate a course element focused on pedagogical issues, specifically, by conveying some of the underlying lesson preparation. For example, at the end of a set of mathematical tasks, we may spend time discussing: 1) how we selected and set up a mathematical task; 2) how we supported the teacher-learners' exploration of the task; and 3) how we shared and discussed the task (Romagnano, Evans & Gilmore, 2008). Finally, a limitation of the study reported here is its reliance on self-report data. In professional development programs such as this one, we encourage researchers to collect data in participants' classrooms in order to assess the program's tangible benefits.

Online courses, sometimes criticized as being merely electronic bulletin boards, have gained currency in recent years, especially for geographically isolated rural students and teachers. At the same time, the mathematics education community has increased its understanding of how participation in a community of practice can support the learning of mathematics. The preliminary results reported here suggest that online professional development environments can capture elements of a community of practice, engage mathematics teachers in meaningful learning experiences, and raise important questions about mathematics and mathematics teaching.

References

Ball, D. L., & Bass, H. (2003). Toward a practice-based theory of mathematical knowledge for teaching. In B. Davis & E. Simmt (Eds.), *Proceedings of the 2002 annual meeting of the Canadian Mathematics Education Study Group* (pp. 3-14). Edmonton, AB: CMESG/GCEDM.

Ball, D. L., & Cohen, D. K. (1999a). Developing practice, developing practitioners: Toward a practice-based theory of professional education. In G. Sykes & L. Darling-Hammond (Eds.), *Teaching as the learning profession: Handbook of policy and practice* (pp. 3-32). San Francisco: Jossey Bass.

Ball, D. L., & Cohen, D. K. (1999b). *Challenges of improving instruction: A view from the classroom.* Washington, DC: Council for Basic Education.

Greeno, J. G. (2003). Situative research relevant to standards for school mathematics. In J. Kilpatrick, W. G. Martin, & D. Schifter (Eds.), *A research companion to principles and standards for school mathematics* (pp. 304-332). Reston, VA: National Council of Teachers of Mathematics.

Mewborn, D. S. (2003). Teaching, teachers' knowledge, and their professional development. In J. Kilpatrick, W. G. Martin, & D. Schifter (Eds.), *A research companion to principles and standards for school mathematics* (pp. 45-52). Reston, VA: National Council of Teachers of Mathematics.

Peressini, D., Borko, H., Romagnano, L., Knuth, E., & Willis, C. (2004). A conceptual framework for learning to teach secondary mathematics. *Educational Studies in Mathematics, 56*, 67-96.

Reeves, C. (2003). *Implementing the No Child Left Behind Act: Implications for rural schools and districts.* Naperville, IL: North Central Regional Educational Laboratory.

Romagnano, L. (1994). *Wrestling with change: The dilemmas of teaching real mathematics.* Portsmouth, NH: Heinemann.

Romagnano, L., Evans, B., & Gilmore, D. (2008). Using video cases to engage prospective secondary mathematics teachers in lesson analysis. In M. S. Smith & S. N. Friel (Eds.), *Cases in mathematics teacher education: Tools for developing knowledge needed for teaching (AMTE Monograph 4)* (pp. 103-115). San Diego, CA: Association of Mathematics Teacher Educators.

Sowder, M. (2005). Challenges and opportunities of peer mentoring: A case study of mentor-mentee's conversations on mentee's teaching. *New England Mathematics Journal*, *37*(2), 50-59.

Stigler and Hiebert. (1999). *The teaching gap*. New York: The Free Press.

U. S. Congress (2001). No Child Left Behind Act of 2001. Public Law 107-110. 107th Congress. Washington DC: Government Printing Office.

---

[i] This program and research are supported by No Child Left Behind Act of 2001, Improving Teacher Quality, Title II: "Metro's Mathematics for Rural Schools Program: Creating Highly Qualified Mathematics Teachers in Rural Schools," Colorado Commission on Higher Education grant CFDA# 84.367B06-07-6. The authors would like to thank Metro's Mathematics for Rural Schools Program's co-designers Jim Loats, Don Gilmore, and Patricia McKenna.

[ii] For more information about the *Elluminate*® software used in Metro's Program, please contact the authors.

Brooke Evans is an Assistant Professor of Mathematical Sciences at the Metropolitan State College of Denver. She teaches and advises students in mathematics and mathematics education and conducts research on mathematics teaching and learning. She is the PI for Metro's Mathematics for Rural Schools Program.

Hamilton Bean is a doctoral candidate in the Department of Communication at the University of Colorado at Boulder. His research centers on the interconnections among organizational communication, discourse, and technology.

Lew Romagnano is Professor of Mathematical Sciences at the Metropolitan State College of Denver, where he works primarily with prospective elementary and secondary mathematics teachers. His scholarly interests include mathematics knowledge for teaching, mathematics assessment, and the learning to teach trajectory.

Chval, K., Lannin, J., and Bowzer, A.
AMTE Monograph 5
Inquiry into Mathematics Teacher Education
©2008, pp. 35-45

4

The Task Design Framework: Considering Multiple Perspectives in an Effective Learning Environment for Elementary Preservice Teachers

Kathryn B. Chval
John K. Lannin
University of Missouri

Angela D. Bowzer
Westminster College

Course design for prospective teachers (PSTs) should be research-based, related to actual instructional practice, and problem-based—involving situations that bring out the complexity of decision making in the mathematics classroom. As our group at the University of Missouri redesigned courses for elementary teachers, we developed a framework for task design to guide our instructional practice. This chapter describes our task design framework that can be used to create, implement, and reflect on instructional tasks that develop PST content and pedagogical knowledge in mathematics content and methods courses.

Mathematics teacher preparation in the United States is highly variable in terms of content, complexity, length, and structure (Floden & Philipp, 2003), often resulting in considerable inconsistencies in the preparation of elementary, middle, and secondary mathematics teachers. Differences in goals, philosophies, and institutional structures (e.g., private vs. public, small vs. large higher education institutions) lead to programmatic-level variability across institutions (Wilson, Floden, & Ferrini-Mundy, 2001). Further variability exists at the course level, leading to disparities in the enactment of programmatic goals and philosophies. These variabilities are problematic because they result in marked differences in the knowledge that preservice teachers (PSTs) develop as part of their mathematics content and/or methods courses.

The nature of the work of mathematics teacher educators contributes to this variability. Mathematics teacher educators often work in isolation and may fail to apply a research-based design process to the development of content and methods courses. Too often, the course design process involves simply selecting a textbook and designing a syllabus based on the textbook. To address these problems, mathematics teacher educators (e.g., Hiebert, Morris, Burke, & Jansen, 2007; Philipp, et al., 2007) have begun to develop frameworks and principles that can be used to guide the design and enactment of content and methods courses for PSTs. Although these frameworks and principles have moved the field forward to an extent, more explicit attention and discussion must be directed toward the design and implementation of courses. This discussion must include the utilization of a research-based design process for designing and implementing courses for PSTs. Despite the emphasis on research-based design for mathematics curriculum at the K-12 level (e.g., Clements, 2007), little discussion has occurred at a national level related to using such a design process in developing mathematics methods courses or mathematics content courses for PSTs.

Furthermore, despite common goals for improving the mathematical and pedagogical knowledge of PSTs, the broader mathematics teacher educator community has limited opportunities for sharing and revising course design features and specific tasks (cf., Taylor & Ronau, 2006). We argue that a focus on instructional tasks is an appropriate initial mechanism for supporting discussion among mathematics teacher educators who seek to improve the design of methods courses. Thus, the purpose of this chapter is to describe a design framework that can be used to create, implement, and reflect on instructional tasks that develop PST content and pedagogical knowledge.

Situating Our Task Design Process

In Fall 2006, we implemented a newly designed two-course elementary mathematics content/methods sequence. This new sequence of courses was a transition for us in two ways: 1) from teaching elementary mathematics methods to teaching mathematics content and methods in the same course; and 2) from a one semester course requirement to a two-course sequence over the academic year. As we began the re-design process, we realized that we needed a theoretical perspective to guide our design process for these courses as we articulate in the following paragraphs.

Our design process utilized features that characterize developmental research using teaching experiments (Cobb, 2000) and curriculum development models (Battista & Clements, 2000). Following Cobb's instructional model, we constructed mathematical and pedagogical learning goals to inform the design of our methods/content courses. To meet the incoming needs of our students, we varied our framework to incorporate components of Realistic Mathematics Education (Gravemeijer, 1995) by framing mathematical tasks in situations that are realistic for preservice classroom teachers, yet embody our Task Design Framework. In the design model for our tasks, we applied the processes of teaching, model building, and hypothesis testing to examine PSTs' learning (Battista & Clements, 2000).

To inform our design process, we collected data from a variety of sources: videotapes from all class sessions, audio-recordings of select group discussions in each class, multiple PST interviews, pre- and posttests related to mathematical knowledge for teaching (Hill, Schilling, & Ball, 2004), and PSTs' work (written work and online discussions). In addition, we recorded course design meetings that occurred prior to and following each class session. However, the purpose of this chapter is not to report findings from this study, but to provide a framework based on the data from a larger study. In the following section, we present a typical task that could be used in a mathematics content or methods course for elementary PSTs. Following this task, we discuss our Task Design Framework and share a revised version of the task.

The Original Task

The task in Figure 1 is a typical task contained in texts for mathematics content or methods courses for elementary PSTs. The task is designed to engage PSTs in the mathematical content of generalizing and justifying generalizations.

> Mary notices that:
> 6×6 is one more than 5×7
> 8×8 is one more than 7×9
> 3×3 is one more than 2×4
>
> Show that for any n in the set of whole numbers, $n \times n$ is one more than $(n - 1)(n + 1)$.

Figure 1. The original task.

Task Design Framework

As we designed tasks, we used a framework related to effective learning environments (Bransford, Brown, & Cocking, 2000) as an overall structure, concentrating on tasks that simultaneously address the following perspectives.

Community: The instructor facilitates the development of a community of learners that values the search for meaning and understanding, builds collaborative relationships, and enhances participation in educational research and practice.
Learner: The instructor identifies and builds on PSTs' knowledge, skills, attitudes, and beliefs as well as strengths, interests, and needs.
Assessment: The instructor continuously monitors and assesses PSTs' progress, encourages PST metacognition, and provides feedback and opportunities for revision.
Knowledge: The instructor selects course content to help PSTs develop the knowledge, competencies, and dispositions necessary for the effective teaching of mathematics. Specifically, elementary PSTs need to know:

- a substantial amount of mathematics content that relates directly to the mathematics content they will teach;
- effective teaching strategies;
- the scope and sequence of mathematics curriculum;
- the general characteristics of elementary students;
- typical student errors and misconceptions, and how to deal with them;
- how to engage in critical reflection and evaluation of one's own teaching; and
- how to work in diverse school settings, including an understanding of the conditions of their students' lives and of the social factors that affect schooling (Ball, Hill, & Bass, 2005; Ma, 1999; Shulman, 1986).

Moreover, we recognized that we needed to design tasks that purposefully addressed these four perspectives. Henningsen and Stein (1997) stated, "The nature of tasks can potentially influence and structure the way students think and can serve to limit or to broaden students' views of the subject matter with which they are engaged" (p. 546). Though Henningsen and Stein focus mainly on student subject matter knowledge, we argue that instructional tasks addressing these four perspectives are essential for the preparation of PSTs. We, as mathematics teacher educators, recognize the importance of addressing the knowledge-centered perspective within university coursework. It is also critical to understand and address the incoming attitudes and beliefs of PSTs (learner-centered perspective) in order to build collaborative relationships and value their search for understanding (community-centered perspective). In order to attend to these perspectives while creating tasks, we considered the research literature related to the beliefs and perceptions that elementary PSTs bring to undergraduate coursework.

PSTs believe that *teaching* is a process of transmitting knowledge, presenting information, or explaining clearly; that it should be rigidly sequenced and structured; it is straightforward, and it is easy. At the same time, they believe that *learning* involves absorbing or memorizing information, listening, giving back what the teacher says, and practicing skills (Ball & McDiarmid, 1987; Calderhead & Robson, 1991; Cohen, 1988; Cuban, 1984; Feiman-Nemser, McDiarmid, Melnick, & Parker, 1988; Feiman-Nemser & Remillard, 1996; Richardson, 1996). In terms of the *discipline of mathematics*, PSTs view mathematics as an arbitrary collection of discrete rules, facts, vocabulary words, and problems (Ball, 1990; Feiman-Nemser & Remillard, 1996). While PSTs indicate that teaching elementary mathematics will be easy (Chval, 2004) and that they already know enough mathematics to teach in the elementary grades (Feiman-Nemser, McDiarmid, Melnick, & Parker, 1988), many do not believe that they are good at mathematics (Ball, 1990).

In addition, PSTs "enter their coursework assuming they already know what they need to know in order to teach" (Ambrose, 2004, p. 91). Many believe that their college preparation is theoretical and thus unrelated to the experiences they will face in the field (Schoonmaker, 1998) and they assume that one learns to teach through teaching, not from participating in college courses (Feiman-Nemser, McDiarmid, Melnick, & Parker, 1988). Additionally, their ideas may be inconsistent with the ideas they encounter in their teacher education courses (Salisbury-Glennon & Stevens, 1999). The research literature suggests that mathematics teacher educators do not have credibility with PSTs. Thus, it is critical to confront the perceptions related to teaching, learning, mathematics, and teacher preparation by inducing uncertainty (Zaslavsky, 2005) through the use of tasks.

Based on this knowledge about incoming PST beliefs and our focus on learning environments, we provide a revised version of the task from Figure 1 in the following section. In addition, we provide excerpts from a few discussions about issues that arose related to the implementation of this task in a content/methods course for elementary PSTs.

The Revised Version of the Task and Class Implementation

We used our Task Design Framework to design the *Double Compare* task (see Figure 2) to engage PSTs in the mathematical content of generalizing and justifying generalizations (knowledge-centered). We introduced the task in the context of a game played by third grade students using a modification of an activity from the *Investigations in Number, Data, and Space®* curriculum (Kliman, Russell, Wright, & Mokros, 1998). The initial conjecture is framed using language a third grade student would use to describe the generalization (learner-centered). The task is intended to provoke uncertainty (Zaslavsky, 2005) in the minds of PSTs regarding the general nature of the statement.

Your third grade students have been playing a version of "double compare" that involves multiplication. Each student chooses two cards, multiplies the numbers on the cards, and compares the result to the results of other students. Michelle noticed that when someone has the "same cards" (e.g., 4 and 4) that the result is always one more than a student who has cards that are one more and one less than the "same cards" (e.g., 3 and 5). See the example below.

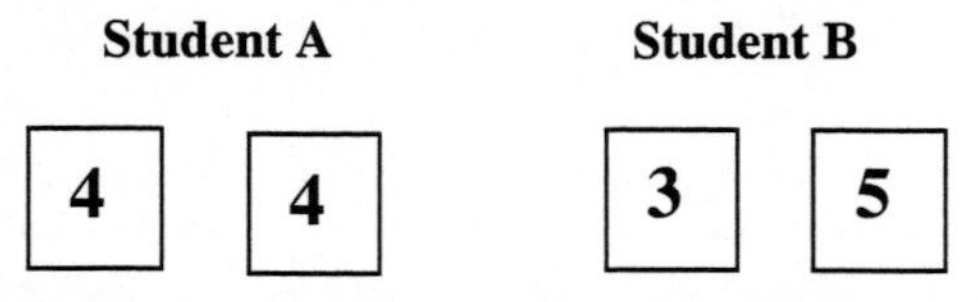

Determine whether this "rule" applies to all whole numbers. Justify your response.

Figure 2. Double compare task.

Below are excerpts from the dialogue that occurred in two groups as they began working on this task. As seen in the dialogue, PSTs struggled to explain what constituted a valid justification for the third grader's claim. The dialogues demonstrate the uncertainty that arose as the PSTs struggled with this task.

Group A

S1: I would say, I mean, if it works for your basic one through ten facts, I would say that it would work for the larger facts as well. But I don't know why that is...

S2: It seems like you're, you're right, if your basic ten (facts), if it worked (for them) then it seemed like it would go all the way up.

The PSTs in Group A discussed the use of an empirical justification (Simon & Blume, 1996) for this situation—attempting to verify a general statement through the use of a finite number of particular instances. As such, they shared the criteria they viewed as acceptable for justifying a generalization, appearing to accept that validating single digit facts served as justification for all whole numbers.

However, as can be seen in the following dialogue, the PSTs in Group B questioned the validity of examining just a few cases and struggled to identify a relationship that explained why the general rule for the task was valid.

Group B

S3: How do you know it's always true though? If you . . .

S4: I don't know why. I'm thinking in terms of groups. Two groups of two.

S3: And three groups of one?

S4: Let's do something besides the one.
[They settle on examining the relationship between the products of 5×5 and 6×4.]

S4: It works... but why?

S5: Well, because you're, okay, if you're taking the number above and the number below you're going to, if you had an odd number then you'll have two even numbers and if you had an even number then you'll have two odd numbers.

Group B extended the discussion of justification beyond examining a few specific cases. They attempted to identify a general relationship that exists in this situation through considering specific cases. During the following whole-class discussion, a member of Group B provided a first attempt at explaining why, through a generic example (Mason, 1996), the conjecture was correct.

S4: Because you're making smaller groups and you're taking away a group and in that group you're distributing what is left in that group into the other groups. Can I show it on the board?

[PST 4 goes to the board.]

S4: So that's 16 and so it's 4 groups of 4 - so the next one would be 3 to 5 is what we're grouping to be one less. So we know this (one item from one group of 4) would go in each group because you're reducing the amount of groups you have ... and the answer's going to be 3 groups of 5. [It] is going to be one less because you'll always have the left over one (showing how distributing one from the group of the four to each of the 3 groups remaining so that there are 5 in each group)... Because there's the same number in this group as how many groups you have (for 4 × 4), and so I'm going from 4 groups to 3 groups, so there's going to be less groups to distribute this into, and so you always distribute one less than what you have. So you'll always have one left over.

This PST explained how removing one group resulted in a leftover object through the use of a particular example. As such she implied that the reasoning underlying this example could be applied to any n by n situation. She began to shape the expectations in the classroom community for the justification of a general statement.

Viewing the Task through the Task Design Framework

In the following paragraphs, we further discuss how the use of the *Double Compare* task connects to the Task Design Framework, examining the design of the task from knowledge-centered, learner-centered, assessment-centered, and community-centered perspectives.

Knowledge-Centered Perspective

The Conference Board of the Mathematical Sciences (2001) recommends that elementary PSTs engage in "representing and justifying general arithmetic claims, using a variety of representations, algebraic notation among them; [as well as] understanding different forms of argument and learning to devise deductive arguments" (p. 20). The *Double Compare* task brings out mathematical knowledge related to justifying general claims and provides an opportunity to discuss what happens when teachers are unsure of the claims made by elementary students in their classrooms. Note that PSTs did not recognize that they could apply algebraic notation. Modifying the task from Figure 1 demonstrates the difficulty that many PSTs have recognizing when to apply algebraic notation when not prompted to do so.

Learner-Centered Perspective

The task builds on PSTs' incoming knowledge of justifying generalizations, allowing them to negotiate what constitutes a valid explanation. Despite the fact that statements of generality and discovering generality are at the very core of mathematical activity (Mason, 1996), many PSTs have experienced traditional K-12 instruction focused on the technique used to construct generalization rather than on the range of the applicability of the technique. The revised task did not suggest a valid justification method, but allowed the PSTs to discuss various methods of justification.

In addition, the task was situated in the context of an elementary classroom. This positioning provides a connection between the task and their future classrooms, demonstrating some of the problems of practice that naturally occur in elementary classrooms.

Assessment-Centered Perspective

The implementation of the task allowed us to assess PSTs' views of generality and the nature of their justifications for specific generalizations. The small-group and whole-class discussions confirmed the strengths and limitations of PSTs' justifications. In addition, the whole-class discussion using the generic example allowed others to self-assess by comparing their justifications to the justification provided by this PST.

Community-Centered Perspective

Lave and Wenger (1991) argue that "[a]ctivities, tasks, functions, and understandings do not exist in isolation; they are part of broader systems of relations in which they have meaning" (p. 53). This task encourages the search for meaning by supporting PSTs as they begin to build a shared understanding of what constitutes an appropriate justification for a general statement. The uncertainty of the task prompted a need for discussing and sharing differing perspectives, which emphasized the importance of creating a collaborative learning environment. In addition, the task allowed for initial defining of norms focused on what is valued regarding such a justification.

Recommendations and Implications

We believe that course design for PSTs should be research-based, related to actual instructional practice, and problem-based—involving situations that bring out the complexity of decision-making in the mathematics classroom. This chapter provides an initial framework for viewing task design for mathematics content and methods courses. The Task Design Framework requires that tasks build on the prior knowledge that PSTs bring to their coursework as well as provide relevant connections to their current and future instructional practice.

Task design can serve as a springboard into deep discussions of the mathematics underlying tasks, as well as connections between crucial mathematical ideas and pedagogical strategies. Task design for mathematics content and methods courses for PSTs must provoke uncertainty about mathematics *and* about how to deal with such mathematical situations in elementary classrooms. Through the implementation of such tasks, we help our PSTs see that their experiences in the teacher preparation program include important ideas directly connected to the challenges they will face in their own instruction. Our use of tasks in this way thus helps establish the credibility of our courses with the PSTs.

Further research is needed related to the use of specific mathematical tasks with PSTs. As a research community of mathematics teacher educators, we know little about the impact of tasks on PSTs' knowledge and beliefs. In addition, despite recent technological advances, we have few venues for sharing the impact of such tasks with the teacher educator community. We must open up our classrooms as research sites in which we share problems of course design and problems of practice so that novice and experienced mathematics teacher educators can establish a common knowledge base that allows for the continued improvement of the preparation of PSTs. Current idiosyncratic methods of course design and implementation are insufficient to meet the political and practical demands of our work.

References

Ambrose, R. (2004). Initiating change in prospective elementary school teachers' orientation to mathematics teaching by building on beliefs. *Journal of Mathematics Teacher Education, 7*, 91-119.

Ball, D. L. (1990). The mathematical understandings that prospective teachers bring to teacher education. *The Elementary School Journal, 90*(4), 449-466.

Ball, D. L., Hill, H. C., & Bass, H. (2005). Knowing mathematics for teaching: Who knows mathematics well enough to teach third grade, and how can we decide? *American Educator, 29*(3) 14-17, 20-22, 43-46.

Ball, D. L., & McDiarmid, G. W. (1987). Understanding how teachers' knowledge changes. *National Center for Teacher Education Colloquy, 1*(1), 9-13.

Battista, M. T., & Clements, D. H. (2000). Mathematics curriculum development as a scientific endeavor. In A. E. Kelly & R. A. Lesh (Eds.), *Handbook of research design in mathematics and science education* (pp. 737-760). Mahwah, NJ: Lawrence Erlbaum Associates.

Bransford, J. D., Brown, A. L., & Cocking, R. R. (Eds.). (2000). *How people learn: Brain, mind, experience, and school.* Washington, DC: National Academy Press.

Calderhead, J., & Robson, M. (1991). Images of teaching: Student teachers' early conceptions of classroom practice. *Teaching and Teacher Education, 7*(1), 1-8.

Chval, K. (2004). Making the complexities of teaching visible for prospective teachers. *Teaching Children Mathematics, 11*(2), 91-96.

Clements, D. H. (2007). Curriculum research: Toward a framework for research-based curricula. *Journal for Research in Mathematics Education, 38*(1), 35-70.

Cobb, P. (2000). Conducting teaching experiments in collaboration with teachers. In A. E. Kelly & R. A. Lesh (Eds.), *Handbook of research design in mathematics and science education* (pp. 307-334). Mahwah, NJ:

Lawrence Erlbaum Associates.
Cohen, D. K. (1988). Teaching practice: Plus a change. In P. Jackson (Ed.), *Contributing to educational change: Perspectives on research and practice* (National Society for the Study of Education Series on Contemporary Issues). Berkeley, CA: McCutchan.
Conference Board of the Mathematical Sciences. (2001). *The mathematical education of teachers.* Providence, R.I. and Washington, D.C.: American Mathematical Society and Mathematical Association of America.
Cuban, L. (1984). *How teachers taught: Constancy and change in American classrooms, 1890-1980.* New York: Longman.
Feiman-Nemser, S., McDiarmid, G. W., Melnick, S., & Parker, M. (1988). *Changing beginning teachers' conceptions: A study of an introductory teacher education course.* East Lansing, MI: National Center for Research on Teacher Education and Department of Teacher Education, Michigan State University.
Feiman-Nemser, S., & Remillard, J. (1996). Perspectives on learning to teach. In F. B. Murray (Ed.), *The teacher educator's handbook* (pp. 63-91). San Francisco: Jossey-Bass Publishers.
Floden, R. E., & Philipp, R. (2003). Report of Working Group 7: Teacher Preparation. In F. K. Lester & J. Ferrini-Mundy (Eds.), *Proceedings of NCTM Research Catalyst Conference* (pp. 171-176). Reston, VA: National Council of Teachers of Mathematics.
Gravemeijer, K. P. E. (1995). *Developing realistic mathematics instruction.* Utrecht, Netherlands: Freudenthal Institute.
Henningsen, M., & Stein, M. K. (1997). Mathematical tasks and student cognition: Classroom-based factors that support and inhibit high-level mathematical thinking and reasoning. *Journal for Research in Mathematics Education, 28*(5), 524-549.
Hiebert, J., Morris, A. K., Burke, D., & Jansen, A. (2007). Preparing teachers to learn from teaching. *Journal of Teacher Education, 58* (1), 47-61.
Hill, H. C., Schilling, S. G., & Ball, D. L. (2004). Developing measures of teachers' mathematics knowledge for teaching. *Elementary School Journal, 105*, 11-30.
Kliman, M., Russell, S. J., Wright, T., & Mokros, J. (1998). *Mathematical thinking at grade 1.* Palo Alto, CA: Dale Seymour Publications.
Lave, J., & Wenger, E. (1991). *Situated learning: Legitimate peripheral participation.* Cambridge, UK: Cambridge University Press.
Ma, L. (1999). *Knowing and teaching elementary mathematics: Teachers' understanding of fundamental mathematics in China and the United States.* Mahwah, N.J.: Lawrence Erlbaum Associates.
Mason, J. (1996). Expressing generality and roots of algebra. In N. Bednarz, C. Kieran, & L. Lee (Eds.), *Approaches to algebra: Perspectives for research and teaching* (pp. 65-86). Dordrecht, The Netherlands: Kluwer Academic Publishers.
Philipp, R. A., Ambrose, R., Lamb, L. L. C., Sowder, J. T., Schappelle, B. P., Sowder, L., et al. (2007). Effects of early field experiences on the

mathematical content knowledge and beliefs of prospective elementary school teachers: An experimental study. *Journal for Research in Mathematics Education, 38*, 438-476.
Richardson, V. (1996). The role of attitudes and beliefs in learning to teach. In J. Sikula, T. Buttery, & E. Guyton (Eds.), *Handbook of research on teacher education* (2nd ed., pp. 102-119). New York: Simon & Shuster Macmillan.
Salisbury-Glennon, J. D., & Stevens, R. J. (1999). Addressing preservice teachers' conceptions of motivation. *Teaching and Teacher Education, 15*, 741-752.
Schoonmaker, F. (1998). Promise and possibility: Learning to teach. *Teachers College Record, 99*, 559-592.
Shulman, L. (1986). Those who understand: Knowledge growth in teaching. *Educational Researcher, 15*(2), 4-14.
Simon, M. S., & Blume, G. W. (1996). Justification in the mathematics classroom: A study of prospective elementary teachers. *Journal of Mathematical Behavior, 15,* 3-31.
Taylor, P. M., & Ronau, R. (2006). Syllabus study: A structured look at mathematics methods courses. *AMTE Connections, 16* (1), 12-15.
Wilson, S. M., Floden, R. E., & Ferrini-Mundy, J. (2001). *Teacher preparation research: Current knowledge, gaps, and recommendations*. Seattle, WA: University of Washington.
Zaslavsky, O. (2005). Seizing the opportunity to create uncertainty in learning mathematics. *Educational Studies in Mathematics, 60,* 297-321.

Kathryn Chval is an Assistant Professor in the Learning, Teaching, and Curriculum Department and Co-Director of the Missouri Center for Mathematics and Science Teacher Education at the University of Missouri. Dr. Chval's research interests include effective preparation models and support structures for teachers across the professional continuum.

John Lannin is an Associate Professor in the Learning, Teaching, and Curriculum Department and Director of Elementary Education at the University of Missouri. Dr. Lannin's research interests include student algebraic reasoning and the development of teacher knowledge.

Angela Bowzer is an Assistant Professor in the Department of Mathematical Sciences at Westminster College in Fulton, MO, where she teaches mathematics methods/content courses for preservice teachers. Her research interests include the relationship between teachers' use of mathematics curricula and their professional identities.

Van Zoest, L. and Stockero, S.
AMTE Monograph 5
Inquiry into Mathematics Teacher Education
©2008, pp. 47-58

5

Concentric Task Sequences: A Model for Advancing Instruction Based on Student Thinking

Laura R. Van Zoest
Western Michigan University

Shari L. Stockero
Michigan Technological University

We describe an approach we have used in our secondary mathematics methods course—what we call concentric task sequences—and share some thinking behind the approach and ways it has supported preservice teachers' development. At the center of the sequence is a carefully chosen mathematical task. Other components of the sequence include watching video of students engaging with the task, and planning and teaching a lesson to students using it. We conclude by reflecting on how our collaboration to understand our preservice teachers' learning better has served as professional development for us as teacher educators.

There is wide agreement about the complexity of teaching mathematics and the difficulty of learning to do it well (e.g., Ball & Cohen, 1999; Borko & Putnam, 1996; National Council of Teachers of Mathematics [NCTM], 2000). Most secondary mathematics teacher education programs have at least one course dedicated to mathematics-specific methodology and some have as many as three or four such courses. A study of syllabi for these courses, however, showed little consistency in the content they contain or even in their goals (Ronau & Taylor, 2006). In the interest of contributing to dialogue about what could, and perhaps should, be in a secondary mathematics methods course, we describe a particular approach we have used in our course—what we call concentric task sequences—and share some thinking behind this approach and ways it has supported preservice teachers' development. We conclude by reflecting on how our collaboration has served as professional development for us as teacher educators.

Context

The course[i] that we focus on here is the first of a sequence of three mathematics methods courses offered to prospective secondary school mathematics teachers at our university. The main focus of this NCTM *Standards*-based course (NCTM, 2000) is on teaching for student understanding by accessing and building on student thinking. Course assumptions include: teaching mathematics is complex and requires a deep understanding of mathematics; and inquiry and analysis are necessary to improve practice. Decisions about course content are made based on our desire to maintain coherence, ground the work in practice, develop mathematical knowledge for teaching, and respect preservice teachers' developmental trajectories.

We have maintained coherence by building the course around a coherent video case curriculum, *Learning and Teaching Linear Functions: Video Cases for Mathematics Professional Development, 6-10* [LTLF] (Seago, Mumme, & Branca, 2004) (see Van Zoest & Stockero, 2008, for details). The classroom video clips in the LTLF curriculum and course-based field experiences in local middle schools ground the work of the course in practice. We draw on what is known about practice-based professional development (e.g. Ball & Cohen, 1999) and purposefully look for ways to use the specifics of practice to ground discussions about generalities. Our emphasis on developing mathematical knowledge for teaching is guided by Ball, Bass, and Hill's (2004) set of pedagogical abilities essential to effective mathematics instruction. We particularly focus on providing opportunities for preservice teachers to develop their abilities to pose good mathematical questions and tasks, design accurate and useful mathematical explanations, represent ideas carefully and translate among mathematical representations, interpret students' ideas, and respond productively to students' mathematical questions. Finally, we extend Simon and Tzur's (2004) set of design principles for planning mathematics instruction to planning instruction for preservice mathematics teachers. That is, we take into account our preservice teachers' current understanding of mathematics and mathematics teaching as we identify the mathematics-specific pedagogical understandings and abilities that we want them to develop. We then use those learning goals as the basis for creating hypothetical learning trajectories, which are grounded in particular sets of learning tasks that make up our Concentric Task Sequence Model.

Concentric Task Sequence Model

We approach moving preservice teachers toward our goal of planning and implementing instruction based on student thinking and understanding through a task sequence conceptualized as seven concentric circles with a rich mathematical task at the center (see Figure 1). The concentric nature of the model indicates that each task explicitly integrates the knowledge gained from previous tasks, and that all tasks in the sequence are built around a single core mathematical task.

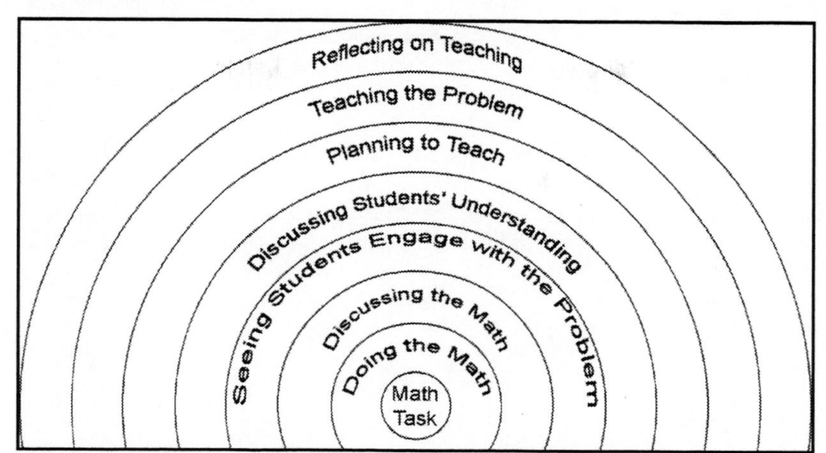

Figure 1: Conceptualization of concentric task sequence model.

Research from the QUASAR project has identified a positive relationship between the use of mathematics tasks that require a high level of cognitive demand and students' mathematical gains; it has also documented the difficulty teachers have implementing such tasks well (Stein & Lane, 1996; Stein, Smith, Henningsen & Silver, 2000). This work has influenced both the mathematical tasks at the core of the model and the pedagogical tasks that we use to support preservice teachers' implementation of tasks that require a high level of cognitive demand with students. In the following sections, we describe the Concentric Task Sequence Model, including an elaboration of each phase in the sequence and examples to illustrate the type of learning each phase supports.

Our Goal for the Sequence

The generation of a hypothetical learning trajectory begins with the identification of a learning goal based on students' current understandings (Simon & Tzur, 2004). For the task sequence we focus on here, we narrowly define our goal as preparing preservice teachers (PSTs) to plan and implement instruction of a particular mathematical problem, based on student thinking and understanding in the moment. More broadly, our goal is to prepare these future teachers to plan and implement other mathematical tasks in a similar manner. As we implement the task sequence, we utilize information we have gained about PSTs' individual understandings—from their written work and participation in class discussions—to facilitate both individual and whole-class development. The complexity of the mathematical and pedagogical tasks in the concentric task sequence provides access points for PSTs with a variety of levels of mathematical knowledge, experience, maturity, and motivation.

At the Center of the Concentric Circles: The Mathematical Task

The task sequence that we use in the methods course is contingent upon a carefully chosen mathematical task, as it becomes the core of all the activities in the sequence. The task on which we base our example, Regina's Logo (see

Figure 2), is drawn directly from the LTLF video curriculum. The task sequence is not dependent on this particular task; there are other tasks with accompanying video clips that would work equally well (e.g., Schifter, Russell, & Bastable, 2002, at the elementary level). What is important is that the task allows students—both the PSTs and the students with whom they work—to think about the mathematics in the problem in multiple ways, leading to a variety of solution strategies and rich mathematical discussions. It is essential that the task requires a high level of cognitive demand (Stein & Smith, 1998); a task that is based on algorithmic knowledge and that most students would think about in only one way would not prompt the type of thinking with which we want to engage our PSTs, and thus would not support subsequent phases in the sequence.

Assume the pattern continues to grow in the same manner. Find a rule or formula to determine the number of tiles in a figure of any size.

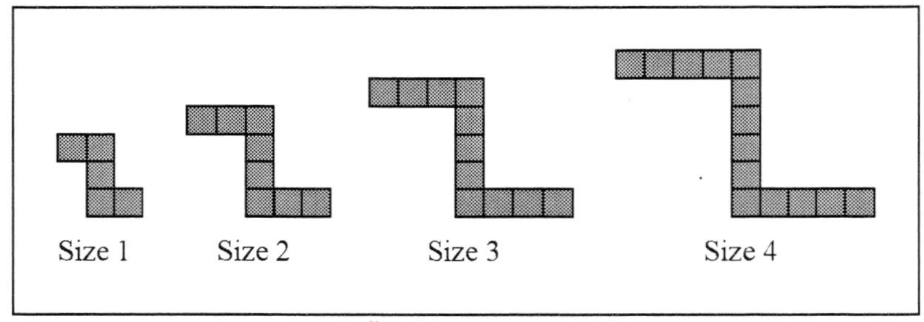

Figure 2. Regina's Logo task.[ii]

The First Phase: Doing the Mathematics

The first phase in which the PSTs engage is solving the mathematical task themselves, making notations in their notebooks regarding their initial solution to the task, as well as any additional solutions that they can generate. Our intent is to elicit their thinking as well as begin to stretch their thinking to consider how others might solve the problem. Initially, many of the preservice teachers have difficulty generating multiple solution strategies and tend to view *their* way of solving the problem as *the* way of solving the problem.

In the case of Regina's Logo, most preservice teachers either use a table of values or visual methods. Those who use a table generally record the total number of tiles for each size and look for patterns in the data. Some notice that the number of tiles increases by three each time, while others simply reduce the pattern to a set of ordered pairs and use an algorithm to find a linear equation for the total number of tiles. Those who use visual methods tend to see the problem in one of four ways: three equal segments of size n, plus the two corners; equal top and bottom segments of size $n + 1$, with a middle segment of size n; equal left and right segments of size n, with a vertical middle segment of size $n + 2$; or a double counting method in which the top and bottom segments each contain $n + 1$ tiles, the middle contains $n + 2$ tiles, and the two corners which have been double-counted are subtracted. Regardless of the solution method they use, almost all the preservice teachers simplify their expression to the form $3n + 2$.

The Second Phase: Discussing the Mathematics

After generating their own solutions to the mathematical task, the PSTs share and discuss their thinking to challenge and deepen their own understanding of the mathematics and to begin to consider the variety of ways others might think about the task. It is often necessary to push the preservice teachers to move from reporting what they did, or what their "answer" was, to discussing how they *thought* about the problem. Our PSTs typically do not recognize how, or even that, their thinking is different from that of others, especially if they arrive at the same final solution. Discussing the mathematics in this way helps move preservice teachers from a narrow, procedural view of mathematics in which there is one "correct" solution towards an understanding that students might think about a problem in one of many mathematically correct ways. This understanding is an important component of the type of teaching we envision—one in which the teacher uses student thinking as the basis for instructional decisions—but comes as a surprise to many PSTs who previously had assumed that everyone thought about mathematical problems in the same way.

As an example, consider Angela's reaction to hearing classmates present their solutions:

> I guess at first I didn't understand. When I looked at it to see if there was a common difference between 1 and 2, 2 and 3, I noticed that it was three, so when everyone started presenting, I was kind of confused at first to like the middle and sides. It was like new ideas.

In this case, Angela had not considered a visual method of solving the problem, and thus, had difficulty understanding her classmates' explanations. If unaddressed, this inability to understand others' thinking could later hinder her ability to facilitate the problem with students. We also have found that such discussions help the preservice teachers realize that an ability to solve mathematical problems does not necessarily imply the kind of deep mathematical understanding that is required to access, evaluate, and build on student thinking.

The Third Phase: Seeing Students Engage with the Problem

The preservice teachers next watch a group of school students—via the LTLF video clips—discuss the same mathematical problem. This gives them an opportunity to check their ideas about how students might think about the problem and to compare students' thinking to their own. In the case of Regina's Logo, the LTLF curriculum contains two video-taped episodes filmed in a 9th grade classroom. The first clip has two students at the board leading a whole-class discussion and recording their classmates' solutions; the second clip shows one particular student, Reymond, presenting his own solution at the board. This second clip includes the teacher questioning Reymond and other students to assess their understanding of the mathematics. In the course of this questioning, another student offers an explanation that is quite difficult to understand; this explanation provides an opportunity to analyze student thinking carefully. The focus on student solution methods in the LTLF videos also provides preservice teachers with what, for most, is an alternate vision of a mathematics classroom and, thus, begins to challenge their preconceptions about what it means to teach and learn mathematics.

The Fourth Phase: Discussing Students' Understanding

Our discussions of the LTLF teaching episodes focus on understanding how the students in the video were thinking about the problem and what the teacher did to either help or hinder that thinking. We ask the PSTs to identify both mathematically and pedagogically important moments and continually push them to support their analyses with evidence from the video transcripts. Our intent is that by discussing these students' thinking, the PSTs will learn how to think about student responses in the moment, rather than dismissing ideas that are not expected or understood. This ability will enable them to respond appropriately to student comments in their field experiences. It is not possible to predict all student responses; however, when PSTs have a core of common strategies and ways of thinking about the problem they are able to focus on those few solutions that are truly novel. These discussions are supported by course readings (e.g., Lannin, Barker, & Townsend, 2006; Stein & Smith, 1998; Wood, 1998) that are assigned at strategic points during the use of the LTLF curriculum and are revisited during the field experience portion of the course. We have found that including such readings is essential to the learning that takes place, as they introduce new ideas, support key ideas discussed in class, and provide a common language that we can use to talk about teaching and learning mathematics.

The following excerpt illustrates how class discussions around the LTLF videos help the preservice teachers carefully consider the reasoning behind students' responses. They have been asked to respond to how Reymond, a student in the video, explained his thinking about the Regina's Logo task.

Vince: I guess when they knew that $x + 1$ plus $x + 1$ gives you $2(x + 1)$, but like couldn't go from $2(x + 1) + x$, to $3x + 2$. Simplify it like each time.

Instructor: So you're questioning whether Reymond could go further with it?

Vince: Well, he didn't simplify it all the way, he just simplified half of it.

Instructor: Ok. So you're saying he recognized that he could multiply two times the quantity but he couldn't completely simplify the equation?

Vince: Yeah.

Randy: I think he left it in that form because it was easier to see how the equation related to the picture. I don't think he was just trying to simplify it down to base form. He kind of keeps it so that everyone could see how he got each part.

Instructor: Is simplifying important? [long pause] In other words, would Reymond's answer have been better if he simplified it?

Thomas: I think being able to simplify is important, but if he'd simplified it to $3x + 2$, I don't see how he would have explained it in a visual way using that equation. He had broken it up into the two different parts so he could say that this part came from here and this part came from here. But if he combined them, I don't know if you can even break up that and look at it visually, $3x + 2$.

Theresa: At this point where they're just trying to recognize patterns, I don't really think it's very important; I think it's important that he could show how he got the equation that he got, that it's more obvious in the more drawn out form.

Vince's assumptions about Reymond's abilities seem to be based on his view of what is mathematically correct and how he would write the final answer. Instead of thinking about the reasons that Reymond may have left his answer as he did, Vince concluded that Reymond did not know how to simplify algebraic expressions. Randy, Thomas, and Theresa, however, offered alternate explanations and provided support for the way that Reymond reported his solution. Responses such as these developed over time as the PSTs were exposed to alternate solution strategies, read mathematics education articles that introduced them to new ideas and language, and were challenged to see what was right about student responses, rather than only focusing on what was wrong.

The Fifth Phase: Planning to Teach the Problem

Planning to facilitate the Regina's Logo problem with groups of 3-5 middle school students in the course field experience is a valuable next step in the PSTs' process of learning to teach. To support their planning, we provide them with a skeletal lesson plan that identifies objectives for the students, objectives for them as the teacher, and prerequisite and subsequent mathematics concepts. The preservice teachers are asked to think about and articulate how they will launch the lesson, what they anticipate might happen as students explore the problem, and how they will summarize the lesson. In the student exploration portion of the plan, they are asked to list learning activities, potential student reactions, and their responses to these student reactions. The intent is to help them carefully think about how students might approach the problem, how they could effectively respond to students, and how they can productively facilitate student learning without removing the challenge of the task. This is perhaps the most challenging part of the task sequence for the preservice teachers. Despite all the previous efforts, it is still natural for them to revert to conceptions from their own school experiences. For example, Chris identified three potential student reactions to his launch of the Regina's Logo task: 1) kicking out an equation because they have done similar problems before; 2) making a table; and 3) making a graph. Although his launch of the problem had left students open to visual solutions—a main emphasis of our class discussions—in conceptualizing possible student approaches, Chris reverted back to the standard equation, table, and graph.

He did, however, address some concepts we talked about in class in the "your response" portion of his plan. In response to the "kicking out an equation" potential student reaction, for example, he listed a series of questions to ask: "How did you get your equation from the picture? Where is each part of your equation represented in the picture? What is changing each time and what is staying the same?" We hypothesize that this is part of Chris' developmental trajectory as he struggles to adopt a new view of teaching and learning mathematics. Although not an ideal lesson plan from our perspective, it helped Chris think about ideas from class and include them, at least to some degree, in his instruction, thus providing him with a way to connect student thinking, instructional planning, and classroom instruction. Helping the PSTs begin to understand the importance of this connection is a key goal of the planning we have discussed here, as well as of the teaching and reflection that take place in the next two phases of the task sequence.

The Sixth Phase: Teaching the Problem

During the field experiences, our intention is that the PSTs not show the students how to solve the problem, but rather, that they engage the students with the problem and appropriately question students in order to access, make sense of, and build on their thinking. Prior to the field experience, the PSTs are given prompts that they will respond to in a post-teaching reflective paper; these prompts require them to analyze how their lesson plan supported them in maintaining the high level of cognitive demand of the task, how their students thought about the problem, and how they as the teacher helped or hindered such thinking. To provide evidence for their analysis, the preservice teachers audiotape their session, are assigned a peer to document their interactions with students, and collect their students' work (see Van Zoest, 2004, for more details about the field experience). Having a carefully-thought-through lesson plan during their teaching experience helps them remain focused on student thinking and asking good questions to elicit thinking, rather than shifting into a "telling" mode of teaching. Chris, for example, in his lesson plan analysis wrote:

> One good aspect about the lesson plan was the predicting how I should respond to the students. Our lesson plan has us press for justification and explanations, which Stein and Smith [1998] associate as factors for [maintaining] high cognitive demands. One instance when a student created a generalized form, I looked at the lesson plan and remembered to ask them

> how their equation fit with the picture, which I would have totally forgot (sic) to ask.

The Seventh Phase: Reflecting on Teaching

Requiring the preservice teachers to reflect on their own practice, using the prompts previously described, builds on the work they have done in analyzing other teachers' practice via the LTLF videos. They are asked to use evidence from their teaching session to support their analyses and to make connections to course readings; this requires them to revisit these readings and make explicit connections between the ideas they contain and their own teaching practice. By engaging in such reflection, the preservice teachers have the opportunity to consider how their planning and implementation affected student thinking during the lesson. Through this process, they come to identify things they did not know they were doing, hear student comments they may have missed, and think more carefully about each student's understanding. The following excerpt from Candi's reflective paper illustrates this. Note that this excerpt is in reference to Regina's Logo 2, a follow-up problem in which the size number is shifted one figure to the left from the original.

> Ryan came up with a really interesting method a few times that I wish we could have explored. Initially, I forgot to ask about it. He uses the size number as a base and then adds the change to it (references transcript). The second time it was right at the end of the hour and we didn't have time to go into it, but if you look at his purple worksheet (which contained the problem), he has a couple of things that are changing. He multiplies the size number by two and adds size minus one to get the numbers of blocks in each size. It looks like he used the first size logo, made an equation to get the answer for the number of blocks and then modified it for every subsequent logo. I wish that we could have played with this for a few minutes, because it could have shown the students that there are several ways of solving the problem.

The reflection phase of the task sequence allows the PSTs to experience reflection as an important part of teaching and a critical first step towards improving practice. Including a sequence of three field experiences in the course provides them an opportunity to use what they learn to refine their teaching in subsequent field experiences.

Conclusion

We have described the Concentric Task Sequence Model (Figure 1), explained some of the thinking behind it, and illustrated some of the ways each phase in the sequence supports preservice teacher development. Understandings towards which we see our PSTs progressing include:

- Being prepared to use a mathematics problem with students requires more than an ability to solve the problem;
- Not all people think about mathematics problems the same way;
- Students are not wrong just because their solution method is confusing or unexpected;
- There is a relationship between student thinking, instructional planning, and instructional implementation; and
- Inquiry and analysis are necessary to improve practice.

Although we believe our task sequence supports PST development in a number of important ways, using such a highly coherent and integrated approach has drawbacks. Specifically, there is a wide range of school mathematics topics that our preservice teachers don't experience in this way, and the question of how much they transfer their learning in one mathematical context to another remains unanswered. As mentioned earlier, getting the PSTs to maintain a focus on student thinking when faced with real students is difficult. We continue to work on providing scaffolds that support their development as they progress through the concentric task sequences. Finally, focusing on PSTs' thinking is fully as challenging and time-consuming as many teachers have found focusing on school students' thinking to be. In particular, the intensive nature of the field experiences and the PSTs' reflections on them (often 7 or 8 page papers), combined with our large class sizes (typically 24 students), makes the quick turn around time and detailed feedback required a formidable task. This dilemma motivates our continued quest to simplify the process without compromising PSTs' learning. Despite these challenges, we have found the results worthwhile and supportive of our students' development as mathematics teachers.

Professional Development for Teacher Educators

Recent attention to the need for professional development for teacher educators (Sowder, 2007) has caused us to reflect on what we have learned as a result of our experiences analyzing our students' development. Working together to design the concentric task sequences and reflect on the PSTs' work has created an opportunity for us to focus on our preservice teachers' thinking and how best to build on it in order to support their learning. This work is analogous to the emphasis in teacher education on studying instructional materials and artifacts of practice in order to make sense of students' mathematical thinking and understanding so that it can be built on through carefully planned instruction (Darling-Hammond & Bransford, 2005; Sowder, 2007). We strongly encourage everyone who is interested in implementing any of the ideas in this chapter to do so with a colleague, or a group of colleagues, in order to create the kind of opportunity that we have found critical to improving our own practice as teacher educators.

References

Ball, D. L., Bass, H., & Hill, H. C. (2004). *Knowing and using mathematical knowledge in teaching: Learning what matters.* Paper presented at the 12th Annual Conference of the Southern African Association for Research in Mathematics, Science and Technology Education, Durban, South Africa.

Ball, D. L., & Cohen, D. K. (1999). Developing practices, developing practitioners. In L. Darling-Hammond & G. Sykes (Eds.), *Teaching as the learning profession: Handbook of policy and practice* (pp. 3-32). San Francisco: Jossey-Bass.

Borko, H., & Putnam, R. T. (1996). Learning to teach. In D. C. Berliner & R. C. Calfee (Eds.), *Handbook of educational psychology* (pp. 673-708). New York: Simon and Schuster Macmillan.

Darling-Hammond, L., & Bransford, J. (Eds.). (2005). *Preparing teachers for a changing world.* San Francisco: Jossey-Bass.

Lannin, J., Barker, D., & Townsend, B. (2006). Why, why should I justify? *Mathematics Teaching in the Middle School, 11*(9), 437-443.

National Council of Teachers of Mathematics. (2000). *Principles and standards for school mathematics.* Reston, VA: Author.

Ronau, R., & Taylor, P. M. (2006). *A framework for examining mathematics methods courses: Discussing appropriate preparation for prospective mathematics teachers.* Paper presented at the Annual Meeting of the Association of Mathematics Teacher Educators, Tampa, FL.

Seago, N., Mumme, J., & Branca, N. (2004). *Learning and teaching linear functions: Video cases for mathematics professional development, 6-10.* Portsmouth, NH: Heinemann.

Schifter, D., Russell, S. J., & Bastable, V. (2002). *Developing mathematical ideas.* Lebanon, IN: Pearson Learning Group.

Simon, M. A., & Tzur, R. (2004). Explicating the role of mathematical tasks in conceptual learning: An elaboration of the hypothetical learning trajectory. *Mathematical Thinking and Learning, 6*(2), 91-104.

Sowder, J. T. (2007). The mathematical education and development of teachers. In F. K. Lester Jr. (Ed.), *Second handbook of research on mathematics teaching and learning* (pp. 157-223). Charlotte, NC: Information Age Publishing.

Stein, M. K., & Lane, S. (1996). Instructional tasks and the development of student capacity to think and reason: An analysis of the relationship between teaching and learning in a reform mathematics project. *Educational Research and Evaluation, 2*(1), 50-80.

Stein, M. K., & Smith, M. S. (1998). Mathematical tasks as a framework for reflection: From research to practice. *Mathematics Teaching in the Middle School, 3*(4), 268-275.

Stein, M. K., Smith, M. S., Henningsen, M. A., & Silver, E. A. (2000). *Implementing standards-based mathematics instruction: A casebook for professional development.* New York: Teachers College Press.

Van Zoest, L. R. (2004). Preparing for the future: An early field experience that focuses on students' thinking. In T. Watanabe & D. R. Thompson (Eds.), *The work of mathematics teacher educators* (pp. 124-140). San Diego, CA: Association of Mathematics Teacher Educators.

Van Zoest, L. R., & Stockero, S. L. (2008). Using a video-case curriculum to develop preservice teachers' knowledge and skills. In M. S. Smith, & S. N. Friel (Eds.), *Cases in mathematics teacher education: Tools for developing knowledge needed for teaching* (pp. 117-132). San Diego, CA: Association of Mathematics Teacher Educators.

Wood, T. (1998). Alternative patterns of communication in mathematics classes: Funneling or focusing? In H. Steinbring, M. G. Bartolini Bussi, & A. Sierpinska (Eds.), *Language and communication in the mathematics classroom* (pp. 167-178). Reston, VA: National Council of Teachers of Mathematics.

---

[i] While a graduate student, the second author worked with the first in this course; this chapter draws on that work and their ongoing collaboration to refine the course and study the student learning that occurs within it.

[ii] Regina's Logo reprinted with permission from **Learning and Teaching Linear Functions: Video Cases for Mathematics Professional Development** by **Nanette Seago, Judith Mumme, and Nicholas Branca**.

Laura R. Van Zoest is a Professor of Mathematics Education at Western Michigan University specializing in secondary school mathematics teacher education. She is particularly interested in the process of becoming an effective mathematics teacher and ways in which university coursework can best support that process.

Shari L. Stockero is an Assistant Professor of Mathematics Education at Michigan Technological University. Her research focuses on how to best support teacher learning and develop reflective practitioners, both in university preservice teacher education courses and in professional development contexts.

Koellner, K., Schneider, C., Roberts, S., Jacobs, J., and Borko, H.
AMTE Monograph 5
Inquiry into Mathematics Teacher Education
©2008, pp. 59-70

6

Using the Problem-Solving Cycle Model of Professional Development to Support Novice Mathematics Instructional Leaders[i]

Karen Koellner
University of Colorado Denver

Craig Schneider
Sarah Roberts
Jennifer Jacobs
University of Colorado at Boulder

Hilda Borko
Stanford University

This chapter is focused on the ways our research team supported newly designated mathematics instructional leaders (ILs) to conduct the Problem-Solving Cycle (PSC) model of professional development with teachers in their schools. We describe the challenges of helping a district implement site-based mathematics professional development and ways in which our research team worked with the ILs to help them understand and employ components of the PSC model. We document the lessons our research group learned from this experience, including general lessons about working with novice ILs and lessons specific to the PSC model.

Recently there has been a strong push at the federal, state, and district levels for widely expanded professional development (PD) opportunities for teachers. For example, the No Child Left Behind Act (U. S. Congress, 2001) requires that states ensure the availability of "high-quality" PD for all U.S. teachers. One increasingly common approach to address this requirement is to identify instructional leaders (ILs) at school sites who can provide such support for their colleagues. Districts and schools across the United States are hiring ILs (also called specialists, coaches, or mentors) to work with teachers. As PD providers, ILs typically employ an informal and loosely defined protocol for their work with teachers, such as demonstration lessons, co-teaching, and peer coaching (Portin, Alejano, Knapp, & Marzolf, 2006). This type of in-house approach is intended to provide sustainable professional development for teachers on a large scale.

Well-prepared PD providers are critical to ensure the effectiveness of PD programs. Recent data from a large-scale investigation of 88 PD programs (in mathematics, science, and technology education) as part of NSF's Local Systemic Change Through Teacher Enhancement Initiative suggests that the quality of PD programs is strongly related to the skills, background knowledge, and preparation of the PD providers (Banilower, Boyd, Pasley, & Weiss, 2006). Specifically, in the field of mathematics education, there is mounting evidence that on-going support and structured learning opportunities for ILs can lead to substantial gains in students' mathematics achievement (e.g., Knapp et al., 2003). For example, Middleton and his colleagues worked with a group of mathematics ILs who were released from their classroom duties 50% of the time to engage in and lead PD programs. Fostering and supporting these "local experts" over time led to sustained improvement in mathematics achievement (Middleton & Coleman, 2005).

Although cultivating the knowledge base, experience, and leadership skills of PD providers is essential, it is often a missing step in educational reform efforts (Loucks-Horsley, Love, Stiles, Mundry, & Hewson, 2003). As Ball and Cohen (1999) noted, "while the role of the teacher educator is critical to any effort to change the landscape of professional development, it is a role for which few people have any preparation...there is little professional development for professional developers" (p. 28). Zaslavsky and Leikin (2004) similarly argued that a central objective of mathematics PD should be to "interweave the professional growth of mathematics teachers with the growth of mathematics teacher educators" (p. 11).

In this chapter, we describe how our research group formed a partnership with a metropolitan school district and used the Problem-Solving Cycle (PSC) model of mathematics PD (Jacobs et al., 2007; Koellner et al., 2007) to support their newly designated mathematics ILs. We describe the challenges of helping a district implement site-based mathematics PD and ways in which our research team worked with the ILs to help them understand and employ components of the PSC model. We document the lessons our research group learned from this experience, including general lessons about working with novice ILs and lessons specific to the PSC model. By the end of the school year, the ILs reported that, in addition to familiarizing them with the PSC model of PD, our meetings had a substantial impact on their self-perceptions as ILs and on their interest in implementing the PSC in their schools.

Conceptual Framework for the Problem-Solving Cycle Model

Situative perspectives on cognition and learning provide the conceptual framework that guided the design of the PSC model. Scholars within a situative perspective argue that knowing and learning are constructed through participation in the discourse and practices of a community, and are shaped by the physical and social contexts in which they occur (Greeno, 2003; Lave & Wenger, 1991). With respect to PD, situative theorists focus on the importance of creating opportunities for teachers to work together on improving their practice as well as locating these learning opportunities in the everyday practice of teaching (Ball & Cohen, 1999; Putnam & Borko, 2000). We share with many teacher educators the view that constructivist and situative theories are interrelated and that learning involves both construction and enculturation (Cobb, 1994; Driver, Asoko, Leach, Mortimer, & Scott, 1994). Three design principles derived from this framework are central to the PSC model: 1) Establishing a professional learning community; 2) Using video from teachers' own classrooms to provide a meaningful context for learning; and 3) Establishing community around video.

The Problem-Solving Cycle Model

The PSC model of mathematics PD is an iterative, long-term approach to supporting teachers' learning. One iteration of the PSC is a series of three interconnected PD workshops in which teachers share a common mathematical and pedagogical experience, organized around a rich mathematical task. This common experience provides a structure upon which the teachers can build a supportive community that encourages reflection on mathematical understandings, student thinking, and instructional practices. The PSC model is designed to be implemented by a knowledgeable facilitator, who carefully plans and implements each workshop and continually monitors the participating teachers' needs and interests.

During Workshop 1, the facilitator introduces a rich task selected on the basis of the teachers' needs and interests. Teachers collaboratively solve the task, debrief their solution strategies, and develop plans for teaching it to their own students. Teachers share ideas about solving and teaching the PSC problem and create lesson plans tailored to the specific class(es) in which they will teach it. The main goal of this workshop is to help teachers develop the content knowledge necessary for planning and implementing the PSC problem. After Workshop 1, each participant teaches the PSC problem in one of his or her mathematics classes, and the lesson is videotaped.

The facilitator designs two subsequent workshops to focus on the teachers' experiences using the problem in their classrooms, relying heavily on video clips and written student work from their lessons. The major analytical focus of Workshop 2 is the role played by the teacher in implementing the problem. The facilitator selects video clips that can serve as a springboard for exploring topics such as how the teachers introduced the task and managed the classroom discourse. Activities in Workshop 3 are centered on a critical examination of students' mathematical reasoning. Facilitators select video clips and examples of students' written work on the PSC problem, and support discussions around topics such as unexpected methods students used to solve the problem and the ways students explained and justified their ideas.

Initial Research on the Problem-Solving Cycle

Our research team developed the PSC as part of a research and PD program that began in 2003. In the first stage of the program we worked with a group of middle school mathematics teachers to develop and refine the model. In fall 2003, we conducted a series of PD meetings that focused on fostering mathematical content and pedagogical content knowledge, developing norms for viewing and analyzing classroom video, and promoting a community of learners. We conducted the first PSC in spring 2004 and two more iterations during the 2004–2005 academic year. The three iterations used different mathematics problems and focused on different aspects of the teacher's role and students' mathematical reasoning (see Borko, Jacobs, Eiteljorg, & Pittman, 2008 for more details). We utilized a design experiment approach (Design-Based Research Collective, 2003) to study, document, and refine the model. In addition, we created a Facilitator's Guide[ii] to help professional development facilitators learn about the PSC and prepare to implement it with teachers.

Scaling Up: Preparing Instructional Leaders to Facilitate the PSC

The next stage in our research program involved "scaling up" the PSC model by introducing it to a group of middle school mathematics instructional leaders in the Aria school district[iii]. Our research goals were to document the processes involved in preparing ILs to implement the PSC, and to further refine the facilitation materials. We began working with Aria in Fall 2006, the first year they had designated a mathematics teacher in each of their 11 middle schools to be an instructional leader. Aria's interest in developing the leadership skills of their ILs as facilitators of school-based mathematics professional development appeared to overlap neatly with our agenda of preparing and supporting PSC facilitators.

The ILs in Aria were selected by their principals and given responsibility for widely varying roles, including holding monthly "early-release day" (ERD) meetings with the other mathematics teachers in their schools, purchasing textbooks, and in some cases analyzing student assessment data. The district also had a full-time mathematics coordinator, Becki, whose multiple roles included helping teachers learn about and navigate a newly adopted textbook series, *Connected Mathematics Project 2*[iv] (CMP2) (Lappan, Fey, Fitzgerald, Friel, & Phillips, 2006a), creating district-wide mathematics assessments, and supporting the mathematics ILs.

Throughout the 2006-2007 school year, Becki led monthly, full-day PD meetings that the district required all middle school mathematics ILs to attend. As part of the collaboration with our research team, Becki sought volunteers from this group to attend additional full-day PSC meetings with our team; four instructional leaders and their principals agreed. Aria's objective was for Becki and the four ILs to take what they learned from our PSC meetings back to the other ILs in the district (during the district IL meetings) as well as to the teachers in their schools (during the ERD meetings).

PSC Meetings with Aria's Instructional Leaders

We held five, full-day PSC meetings with Aria's mathematics coordinator and four middle school mathematics ILs during the 2006-2007 school year. These meetings were conducted by two members of our research team. At each meeting, at least one other member was present to videotape the session and participate as appropriate. Drawing from our field notes and videotaped records, we briefly describe these meetings, including our goals for the meeting, the major events that took place, and the lessons our research team learned as the meetings unfolded.

PSC Meeting 1: Overview of the PSC Model and Selecting a Task

Meeting goals and activities. The goal for our initial meeting was to introduce the PSC model. To this end, we shared a detailed PowerPoint presentation, providing an overview of the PSC model and examples of our research team's prior experiences working with middle school mathematics teachers using the model. Selecting a rich mathematical task is the first component in implementing the PSC model, and one of the most critical decisions for facilitators. We brought two possible tasks to the meeting and asked the ILs to examine each task and select one to use. After an extended discussion about which task would be most appropriate for their students and teachers, the group decided on the Pool Border problem (Lappan, Fey, Fitzgerald, Friel, & Phillips, 2006b), explaining that it better matched the mathematics content and processes addressed in the CMP2 middle school curriculum.

Lessons learned. Becki came to the first meeting optimistic and enthusiastic about the partnership with our research team, counting on us to help motivate and support the instructional leaders. However, although all of the instructional leaders had chosen to take part in the program in order to foster their leadership skills, they came to the meeting with virtually no information about the nature of our proposed PD agenda. At the beginning of the meeting, the group talked at length about the current state of mathematics PD in Aria. The ILs explained that not only did they have limited experience with and understanding of their role, they were especially concerned about having to implement an unfamiliar PD model with the teachers in their schools, and they did not know how they would manage the time constraints and sometimes competing priorities felt at the school and district levels.

Our research team responded to these concerns by explaining that the PSC model is flexible and extremely responsive to participants' specific and changing needs, particularly with respect to mathematics knowledge, pedagogical practices, community development, and available time for PD. Despite this response, the ILs remained cautious about committing to use the PSC model with the other teachers in their buildings. Over the course of the year, we came to understand more clearly that newly designated ILs may not be ready to dive into any specific PD model at the beginning of their tenure. In addition, as we learned more about each participant's individual circumstances and needs, we gained a better sense of how to support and encourage them to take an active role as an IL and as a facilitator of the PSC.

PSC Meeting 2: Debriefing the School-Site Meeting and Discussing Video

Meeting goals and activities. Our goal for the second meeting was to address the next several phases of the PSC model, after facilitators have selected (and solved) the PSC task. These phases include developing individual lesson plans to implement the selected task, considering logistics associated with videotaping teachers, fostering a community that is comfortable being videotaped and watching video of themselves, and selecting clips from videotaped lessons as a springboard for focused discussions. Shortly after the first PSC meeting, Becki held a district wide IL meeting during which they worked on and discussed the Pool Border task. In addition, Becki had encouraged the ILs to use the task in their classrooms. Two of the ILs from our group had done so; one videotaped her own lesson and the other collected student work samples. Therefore, although we had anticipated spending a large chunk of our second PSC meeting time working with the ILs to develop detailed lesson plans for the Pool Border task, it seemed more prudent to examine and discuss these artifacts. We also talked about videotaping logistics, fostering community around video, and selecting video clips.

Lessons learned. From this meeting, we realized the extent to which activities that take place between PSC meetings can directly impact our intended agenda. It was critical to allow time at the beginning of each PSC meeting to gather information about relevant PD activities at the district and school level that took place following the previous meeting. We also became aware of the importance of communicating with the participants before meetings to learn about interim activities that might influence the nature of the meetings.

PSC Meeting 3: Introducing the Snakes in Snakewood Problem (Roodhardt, Kindt, Burrill, & Spence, 1997)

Meeting goals and activities. For the third meeting we brought a new mathematics task in order to support the ILs' emerging understanding of the student thinking component of the PSC model. In particular, we wanted the ILs to examine the role of video in fostering extended conversations about student reasoning related to the PSC task. The task we brought to the meeting, the Snakes in Snakewood Problem, was one that our research team had used in our initial development of the PSC model; thus, we had video of middle school lessons that we could view with the ILs. The group worked on and discussed the problem and then watched a video clip showing students solving the same problem. The ILs considered aspects of algebraic reasoning, such as various ways the students generalized their patterns, and they talked about discussion questions they could use if they were to facilitate the viewing of this clip.

Lessons learned. One lesson we learned at this stage of our PD work with Aria's ILs was that participation in the PSC as teacher-learners is a critical component in preparing for the facilitation role. We now hypothesize that it takes at least one iteration of experiencing the PSC *as learners* for ILs to grasp the overall structure of the model, and to see firsthand how the different components work together to support teacher learning. After they took part in one (adapted) iteration of the PSC, largely in their role as classroom teachers, the ILs were much more enthusiastic about the model, confident in their ability to facilitate the workshops, and interested in implementing it with teachers in their buildings.

PSC Meetings 4 & 5: Considering the PSC Components from a Facilitator's Perspective

Meeting goals and activities. Our goals for the fourth and fifth meetings were for the ILs to consider the following components of the PSC model from a facilitator's perspective: understanding the characteristics of an appropriate PSC mathematics task; using video to explore the teacher's role in a PSC lesson; and choosing video clips and developing focused discussion questions. In the fourth meeting, we brought a selection of three mathematics tasks for the ILs to solve. We then discussed whether and why each of these tasks would be appropriate for the PSC, and we listed important characteristics of PSC tasks, such as having multiple entry/exit points, eliciting different solution strategies, and the likelihood of fostering mathematically productive discourse. Lastly, the ILs watched a video clip of a middle school teacher implementing the Snakewood problem and considered how they might use such a clip with their own teachers.

In the fifth meeting, we focused on selecting video clips and developing questions to elicit discussion. Unlike prior meetings, where we had pre-selected clips for the ILs to examine, during this meeting the ILs picked out short video clips from a forty-minute excerpt of a teacher's Snakes in Snakewood lesson. They worked in two groups; one group focused on the teacher's role and the other focused on student thinking. Each group watched the video, selected one or more short clips, and developed questions to elicit discussion to match their focus (teacher role or student thinking). The groups then shared their selections, described why they choose the clips, and responded to one another's discussion questions.

Steps forward. During these final two PSC meetings, the ILs talked at length about their increasing comfort with the PSC model and their growing enthusiasm for taking an active role as facilitators of the model. They spoke of their intention to use the overall structure of the model to lead PD efforts with the mathematics teachers in their building during the following school year. Moreover, they appeared determined to take advantage of district- and school-mandated PD time for the 2007-2008 school year to put what they learned as PD providers into practice.

Overall Impacts of the Program on the Instructional Leaders

During our final PSC meeting we conducted individual face-to-face interviews with the ILs, asking them to consider the impact of the PD program on their learning as both teachers and instructional leaders. These interviews were audio-recorded and transcribed. Our research team systematically examined the transcripts, as well as videotaped conversations with the ILs during meetings four and five, in order to document the ILs' impressions of the PD program's major areas of impact.

Consistent with the literature showing that the role of the professional developer is often ill-defined (Portin et al., 2006), the Aria ILs began the PSC meetings with little understanding about the expectations of the district and their schools for sustained mathematics PD and the role they would play in that regard. As one IL, Trina, commented, "I think we came in not knowing what we were doing at all." By the end of the year they had a much clearer understanding of how the PSC could be adapted for use as a PD model with the teachers in their buildings, and they began to take ownership of the process. For example, they told us of their intention to use the mathematics PD time allocated by their administration to implement the PSC during the 2007-08 school year.

The ILs were particularly enthusiastic about working collaboratively with other mathematics teachers to solve rich problems and share solutions. In addition, by deeply analyzing student thinking during the meetings, they gained a strong appreciation of the ways students at different grade levels can engage in the same rich mathematics task. As Lisa explained, "It's really nice to see how other people solve problems because that helps you see how kids are solving problems in different ways too." Janice told us,

> The thing that I took away from this experience in terms of my role as a math teacher was the math problems. Not just doing the problems with the kids, but reflecting on the way that I actually teach them...because as we got back together, and we talked about it and the way different people taught it and that type of thing, it makes you reflect inwardly as a teacher.

The ILs became very interested in videotaping themselves and other teachers implementing PSC tasks in their classrooms. They talked about how powerful it was to watch video clips from PSC lessons and examine the teacher's role and student thinking. Millie noted,

> What stands out [about the PSC meetings] is the value of videotape - watching videotape and being videotaped. The willingness to be videotaped is critical for your own development as well. It gives you a chance to evaluate how you are as a teacher, or how you respond back to kids.

With respect to facilitating the PSC with other teachers in their buildings, she added, "You have convinced me that [videotaping] is powerful. Now I have some ways to show teachers that are powerful also."

Like the ILs, our research team learned valuable lessons from each meeting – lessons that enhanced our thinking about the PSC and steadily improved our delivery of the PD program. Our willingness to take Aria's district, school, and individual ILs' needs into account demonstrated the flexibility of the PSC model and had a very positive impact on the ILs. This impact was noted, for example, by Becki, who reflected, "You looked at all of the nitty-gritty components of what goes on within a specific school district that would prohibit the level of growth and then you adapted. This showed the teachers that you valued them." Several conversations in the final two workshops centered around the importance of valuing the ILs' input regarding specific PD activities and time structures, and tailoring the program to their unique needs. Millie commented,

> Your willingness to find out what our needs were has opened up a lot of doors for us to go back now and work with our individual schools. Because at first when we started this program, it was like, there is no way [this can happen].

Janice added,

> You guys really listen to us. And you brought us your ideas. But, yet, you molded them to something that we could really use.... We always talk about [this experience], going back, as the best opportunity we've had for professional development, where somebody listened to our needs.

Implications for the Field of Mathematics Professional Development

The insights we gained through our experiences using the PSC model to support Aria's novice ILs in their pursuit of learning about and providing site-based mathematics PD add to the emerging literature on the professional development of professional developers. We learned, for example, that it was important for the instructional leaders to experience the PSC model as learners before attempting to implement it as facilitators. It also became clear that establishing a professional learning community and tailoring the PSC model to the Aria context were critical aspects of the success of our work with the ILs. In combination, these design features helped to build the leadership capacity of the novice PD facilitators, and we conjecture that they are essential for the successful "scale-up" of the PSC professional development model. We also hypothesize that they are relevant to the preparation of facilitators for other models of professional development.

We believe that these conjectures regarding the leadership capacity of professional development providers warrant further research. We also encourage researchers in the field to consider developing tools to measure changes in leadership capacity, in order to foster more quantitative investigations. Constructs, such as professional knowledge, professional identity, and community-building skills, will undoubtedly come into play and need to be considered in a more deliberate fashion. Building the body of literature in the emerging field of professional development for PD providers is a critical step in the complex arena of promoting student learning.

References

Ball, D. L., & Cohen, D. K. (1999). Developing practice, developing practitioners: Toward a practice-based theory of professional education. In L. Darling-Hammond & G. Sykes (Eds.), *Teaching as the learning profession: Handbook of policy and practice* (pp. 3-32). San Francisco: Jossey-Bass.

Banilower, E. R., Boyd, S. E., Pasley, J. D., & Weiss, I. R. (2006). *Lessons from a decade of mathematics and science reform: A capstone report for the Local Systemic Change Through Teacher Enhancement Initiative*. Chapel Hill, NC: Horizon Research.

Borko, H., Jacobs, J., Eiteljorg, E., & Pittman, M. E. (2008). Video as a tool for fostering productive discourse in mathematics professional development. *Teaching and Teacher Education, 24,* 417-436.

Cobb, P. (1994). Where is the mind? Constructivist and sociocultural perspectives on mathematical development. *Educational Researcher, 23*(7), 13-20.

Design-Based Research Collective. (2003). Design-based research: An emerging paradigm for educational inquiry. *Educational Researcher, 32*(1), 5-8.

Driver, R., Asoko, H., Leach, J., Mortimer, E., & Scott, P. (1994). Constructing scientific knowledge in the classroom. *Educational Researcher, 23*(7), 5-12.

Greeno, J. G. (2003). Situative research relevant to standards for school mathematics. In J. Kilpatrick, W. G. Martin, & D. Schifter (Eds.), *A research companion to Principles and Standards for School Mathematics* (pp. 304-332). Reston, VA: National Council of Teachers of Mathematics.

Jacobs, J. K., Borko, H., Koellner, K., Schneider, C., Eiteljorg, E., & Roberts, S. A. (2007). The Problem-Solving Cycle: A model of mathematics professional development. *Journal of Mathematics Education Leadership, 10(1),* 42-57.

Koellner, K., Jacobs, J., Borko, H., Schneider, C., Pittman, M., Eiteljorg, E., et al. (2007). The problem-solving cycle: A model to support the development of teachers' professional knowledge. *Mathematical Thinking and Learning, 9(3),* 271-303.

Knapp, M. S., Copland, M. A., Ford, B., Markholt, A., McLaughlin, M. W., Milliken, M., et al. (2003). *Leading for learning sourcebook: Concepts and examples*. Center for the Study of Teaching and Policy. Seattle: University of Washington. Retrieved February 14, 2007, from www.ctpweb.org

Lappan, G., Fey, J. T., Fitzgerald, W. M., Friel, S. N., & Phillips, E. D. (2006). *Connected mathematics* 2. Boston, MA: Pearson Prentice Hall.

Lappan, G., Fey, J. T., Fitzgerald, W. M., Friel, S. N., & Philips, E. D. (2006). *Say it with symbols*. Boston, MA: Pearson Prentice Hall.

Lave, J., & Wenger, E. (1991). *Situated learning: Legitimate peripheral participation.* Cambridge, UK: Cambridge University Press.

Loucks-Horsley, S., Love, N., Stiles, K. E., Mundry, S., & Hewson, P. W. (2003). *Designing professional development for teachers of science and mathematics*. Thousand Oaks, CA: Corwin.

Middleton, J. A., & Coleman, K. (2005). Learner-centered teacher leadership in mathematics education. *Journal of Mathematics Education Leadership*, *8*(1), 25-30.

Portin, B. S., Alejano, C. R., Knapp, M. S., & Marzolf, E. (2006). *Redefining roles, responsibilities, and authority of school leaders*. Center for the Study of Teaching and Policy. Seattle: University of Washington. Retrieved February 14, 2007, from www.ctpweb.org

Putnam, R., & Borko, H. (2000). What do new views of knowledge and thinking have to say about research on teacher learning? *Educational Researcher, 29*(1), 4-15.

Roodhardt, A., Kindt, M., Burrill, G., & Spence, M. S. (1997). Patterns and symbols. In National Center for Research in Mathematical Sciences & The Freudenthal Institute (Eds.), *Mathematics in context*. Chicago: Encyclopedia Britanica.

U. S. Congress (2001). No Child Left Behind Act of 2001. Public Law 107-110. 107th Congress. Washington DC: Government Printing Office.

Zaslavsky, O., & Leikin, R. (2004). Professional development of mathematics teacher educators: Growth through practice. *Journal of Mathematics Teacher Education, 7*, 5-32.

---

[i] The professional development program discussed in this chapter is one component of a larger project entitled *Supporting the Transition from Arithmetic to Algebraic Reasoning* (STAAR). The STAAR Project was supported by NSF Proposal No. 0115609 through the Interagency Educational Research Initiative (IERI). The views shared in this article are ours, and do not necessarily represent those of IERI. We would like to thank the instructional leaders and district mathematics coordinator for working with us. It was a tremendous learning experience for our research team and we are grateful for their time, energy, and commitment to mathematics education.

[ii] A draft of the Facilitator's Guide to Planning and Conducting the Problem-Solving Cycle is available on our website:

http://www.colorado.edu/education/staar/

[iii] The name of the school district and all school and district personnel mentioned in this chapter are pseudonyms.

[iv] For more information on CMP2, see their website: http://www.phschool.com/cmp2/

Karen Koellner is an Associate Professor in Mathematics Education at the University of Colorado Denver. Her primary research interests are in teacher professional learning, students' mathematical thinking, and ways to make mathematics accessible to all students.

Craig Schneider is currently a Senior Instructor/Master Teacher at the University of Colorado at Boulder's CU-Teach program for undergraduates pursuing science and mathematics K-12 teaching. He recently completed a Post-

Doctoral Research Specialist position with the Center of Mathematics Education of Latinos/as at the University of California at Santa Cruz.

Sarah Roberts is a doctoral candidate in Mathematics Curriculum and Instruction at the University of Colorado at Boulder. Her interests include professional development, mathematics knowledge for teaching, and supporting English language learners in the mathematics classroom.

Jennifer Jacobs is a Faculty Research Associate at the University of Colorado at Boulder in the Institute of Cognitive Science. Her primary interests are in educational and comparative research, including instructional practice, mathematics education, and methodological issues.

Hilda Borko is a Professor at Stanford University. Her research explores teacher cognition and the process of learning to teach, with an emphasis on changes in novice and experienced teachers' knowledge and beliefs about teaching, learning, subject matter, and their classroom practices.

Hughes, E., Smith, M., Boston, M., and Hogel, M.
AMTE Monograph 5
Inquiry into Mathematics Teacher Education
©2008, pp. 71-83

7

Case Stories: Supporting Teacher Reflection and Collaboration on the Implementation of Cognitively Challenging Mathematical Tasks[i]

Elizabeth K. Hughes
University of Northern Iowa

Margaret S. Smith
University of Pittsburgh

Melissa Boston
Duquesne University

Michael Hogel
Mt. Lebanon School District

"Case Stories" is a protocol that invites inquiry into and promotes reflection on professional practice through collaborative examination of classroom artifacts (e.g., student work, lesson plans, video). In this chapter, we illustrate how the Case Story process provided a mechanism for secondary mathematics teachers to discuss publicly their teaching. The Case Story process also enabled teachers to receive feedback from like-minded colleagues who were able to raise issues that teachers could continue to reflect upon as they endeavored to refine their practice.

Research shows that mathematical tasks that provide the greatest opportunities for students to think and reason are the most difficult for teachers to implement well during instruction (Stein, Grover, & Henningsen, 1996; Stigler & Hiebert, 2004). All too frequently, students' opportunities for thinking and reasoning are lost as cognitively-challenging tasks decline into procedural exercises that require the application of rules and memorized facts. Factors that contribute to task decline include: routinizing problematic aspects of the task; not holding students accountable for high–level products; shifting the emphasis from meaning, concepts, or understanding to the correctness or completeness of the answer; and providing students with insufficient time to wrestle with the demanding aspects of the task (Henningsen & Stein, 1997).

Although these and other factors can influence the decisions a teacher makes during the lesson, and ultimately impact students' opportunities to learn mathematics, a teacher may be unaware of the ways in which his/her actions and interactions in the classroom impact what students learn and how they learn it. Hence, teachers need an opportunity to reflect on the ways in which their instruction supports or inhibits students' engagement in high-level cognitive processes (Stein, Smith, Henningsen, & Silver, 2000).

Beginning with the work of Schon (1983, 1987), reflection has received attention as an important tool for practitioners. Several studies suggest that reflecting on instructional practice in light of new ideas about teaching and learning can support teachers' development of more inquiry-oriented instruction (e.g., Smith, 2000; Wood, Cobb, & Yackel, 1991). According to Hart (1991), "the process of constructing new knowledge, whether by students or teachers, is facilitated through reflection on the experiences that are motivating the change. Teachers need the opportunity to look back on their teaching strategies – to reflect on the outcome of their behaviors – and to learn from the experience" (p. 80).

Reflection can be fostered by the use of frameworks that make focal specific aspects of instruction and provide a lens through which to examine practice. For example, Stein and her colleagues (Stein & Smith, 1998; Stein et al., 2000) argue that reflecting on instruction through the lens of the Mathematical Tasks Framework (MTF)[ii] provides teachers the opportunity to examine classroom activity critically in terms of the cognitive demands it places on students and to gain insights into how they may have done things differently.

Although reflection and analysis are often individual activities, they can be enhanced through collaboration with respectful colleagues (National Council of Teachers of Mathematics [NCTM], 2000). Stigler and Hiebert (1999) argue that collaborating with colleagues to analyze and discuss teaching and students' thinking is a powerful form of professional development. According to NCTM (2000), "opportunities to reflect on and refine instructional practice – during class and outside of class, alone and with others – are crucial in the vision of school mathematics outlined in *Principles and Standards*" (p. 19).

One methodology for encouraging teacher reflection on practice and collaboration with colleagues is *Case Stories* (Ackerman, Maslin-Ostrowski, & Christensen, 1996). Case Stories is a protocol that invites inquiry into and promotes reflection on professional practice through the collaborative examination of classroom artifacts (e.g., student work, lesson plans, video). Research suggests that Case Stories is an effective professional development tool because it is particularly well-suited to adult learning styles (Maslin-Ostrowski & Ackerman, 1998; Duncan & Clayburn, 1997).

In this chapter we describe a professional development initiative that focused on the selection and enactment of cognitively challenging tasks. We also discuss the use of Case Stories as a vehicle for supporting teachers' inquiry into and reflection on practice and their collaboration with colleagues.

Professional Development that Fosters Learning and Reflection

The professional development described herein was conducted under the auspices of the ESP (Enhancing Secondary Mathematics Teacher Preparation) project intended to support middle and high school mathematics teachers to develop inquiry-oriented instructional practices and to develop as leaders who could mentor beginning teachers. During the first year in the project, teachers attended six sessions focused on mathematics teaching and learning and a full week during the summer focused on becoming a mentor. During the second year in ESP, each teacher, along with the preservice teacher who had been assigned to her classroom, participated in five sessions aimed at developing a shared understanding of how to support students' engagement in cognitively challenging activity by attending to and building on students' mathematical thinking.

The professional development program blended a practice-based approach (Ball & Cohen, 1999; Smith, 2001) with a set of scaffolded field experiences (Borasi & Fonzi, 2002) and Case Story debriefings. During ESP sessions, teachers engaged in a set of practice-based experiences (e.g., analyzing student work samples and narrative and video cases) that focused on selecting and implementing cognitively-challenging tasks (Stein et al., 2000). At the conclusion of each session, teachers were given an assignment that provided an opportunity to "test out" their developing knowledge by applying the ideas explored in the professional development setting to their own classroom instruction. These scaffolded field experiences were intended to sustain and personalize teachers' focus on implementing cognitively challenging tasks in ways that supported students' mathematical thinking. Teachers were asked to return to the next session with written reflections and any evidence or artifacts (i.e., student work, lesson plans, transcribed or paraphrased interactions, video- or audio-taped segments of the lesson) that would help them tell the "story" of how the new ideas "played out" in their classroom.

The Case Story protocol provided a mechanism for teachers to begin talking about teaching and learning in their classrooms, to reflect on teaching and learning and to collect evidence to support their claims, and to help teachers develop a critical stance towards teaching by listening to and respectfully questioning the practice of their colleagues. Case Stories progressed from general descriptions of lessons based on what teachers recalled to more evidenced-based representations of instruction and learning using artifacts generated during the lesson to support claims. Through this process, teachers became comfortable publicly discussing their experiences in implementing challenging mathematical tasks. The Case Stories protocol provided a non-threatening venue for teachers to discuss their implementation of the ideas that emerged from the ESP sessions with colleagues and with members of the professional development team.

The general Case Story protocol used in our work with teachers was adapted from the work of Ackerman, Maslin-Ostrowski, and Christensen (1996) and consisted of four phases, each lasting 5-10 minutes:

1. The story-teller shares the story (e.g., brief description of what occurred during the lesson) or classroom artifacts with a small group of colleagues. The group members review the story and/or artifacts and use a recording sheet (a T-chart marked with the headings "noticing" and "wondering") to make note of what they see in the evidence and the questions it raises for them;
2. The group members make factual statements in the form of "I noticed…" that draw on the presented evidence and refrain from making evaluative comments or statements of personal preference (during this time, the story-teller remains quiet, listens, and takes notes on her own "noticing" and "wondering" T-chart);
3. The group members make statements in the form of "I'm wondering…" that focus on aspects of instruction that appear to be influencing students' opportunities to learn (during this time, the story-teller remains quiet, listens, and takes notes); and
4. The story-teller shares his/her perspective on the lesson, and may respond to the "noticings" and "wonderings" of the group, drawing on the notes (s)he recorded during the earlier phases of the process.

Unlike Lesson Study, where groups of teachers collaboratively plan a lesson and evaluate the lesson taught by a member of the group, Case Stories invite teachers who were not involved in planning or observing the lesson to examine evidence from a lesson and to raise issues for the teacher to consider. Hence, Cases Stories are a flexible mechanism for discussing teaching because they do not require collaborative planning and shared observation. Although this can also be seen as a limitation, in our view, Case Stories can be used in a range of situations and perhaps serve as a precursor to or catalyst for Lesson Study. In both Lesson Study and Case Stories, it is crucial to establish a non-threatening atmosphere in which teachers perceive feedback from colleagues as an opportunity for self-improvement rather than as a personal evaluation (Stigler & Hiebert, 1999). Towards this end, the non-threatening language of "noticing" and "wondering" can be used to promote a thoughtful examination of practice (Smith, in press). [iii]

In the next section, we illustrate the Case Story process and discuss how storytelling through student work can become a mechanism for teachers to inquire into and reflect on instruction.

Case Stories: Storytelling through Student Work

After working for several sessions on identifying cognitively challenging mathematical tasks and analyzing narrative cases for factors that affect the maintenance and decline of a task during implementation (see Stein et al., 2000), we asked teachers to apply these ideas to their own classrooms. Specifically, we asked teachers to: 1) teach a lesson based on a cognitively challenging mathematical task of their choice, 2) select several pieces of work produced by students during the lesson that they felt accurately reflected the lesson, and 3) bring blinded copies of the student work to share at the next session. During the subsequent session, teachers engaged in conversations about their teaching through the four-phase Case Story protocol described earlier. Participants were arranged in small groups of 3-4 teachers. Each teacher took a turn as the storyteller while the others took on the role of inquirers, trying to understand the impact of teaching on student learning.

In the remainder of this section, we focus on a Case Story that was shared by Charles, a mentor teacher who was participating in the second year of ESP. Charles had made progress on identifying and implementing cognitively challenging tasks, but such instruction was still not routine for him. The lesson that was the focus of his Case Story was from an eighth grade class working on algebra and involved the "Concert Tickets Question" task (shown in Figure 1).

Concert Tickets Question

You are so excited about an upcoming concert that you ask all of your friends to join you. The tickets are sold at two different prices, the center section seats cost $10 more than the balcony seats.

You buy 7 center seats and 5 balcony seats for a total of $310.

Please find the cost of each type of seat.

Remember to show all of your work and provide a detailed, written explanation.

Figure 1. "Concert Tickets Question" task from Charles' case story.

The four samples of student work shown in Figure 2 are representative of the eight responses that Charles selected and shared with teachers during the Case Story. All of the work samples used either substitution (as shown in A and B of Figure 2) or elimination (as shown in C and D of Figure 2). In addition, all of the work samples showed procedural explanations of the process used to solve the task, although they varied in the amount of detail provided (e.g., see the differences in explanations provided in C and D).

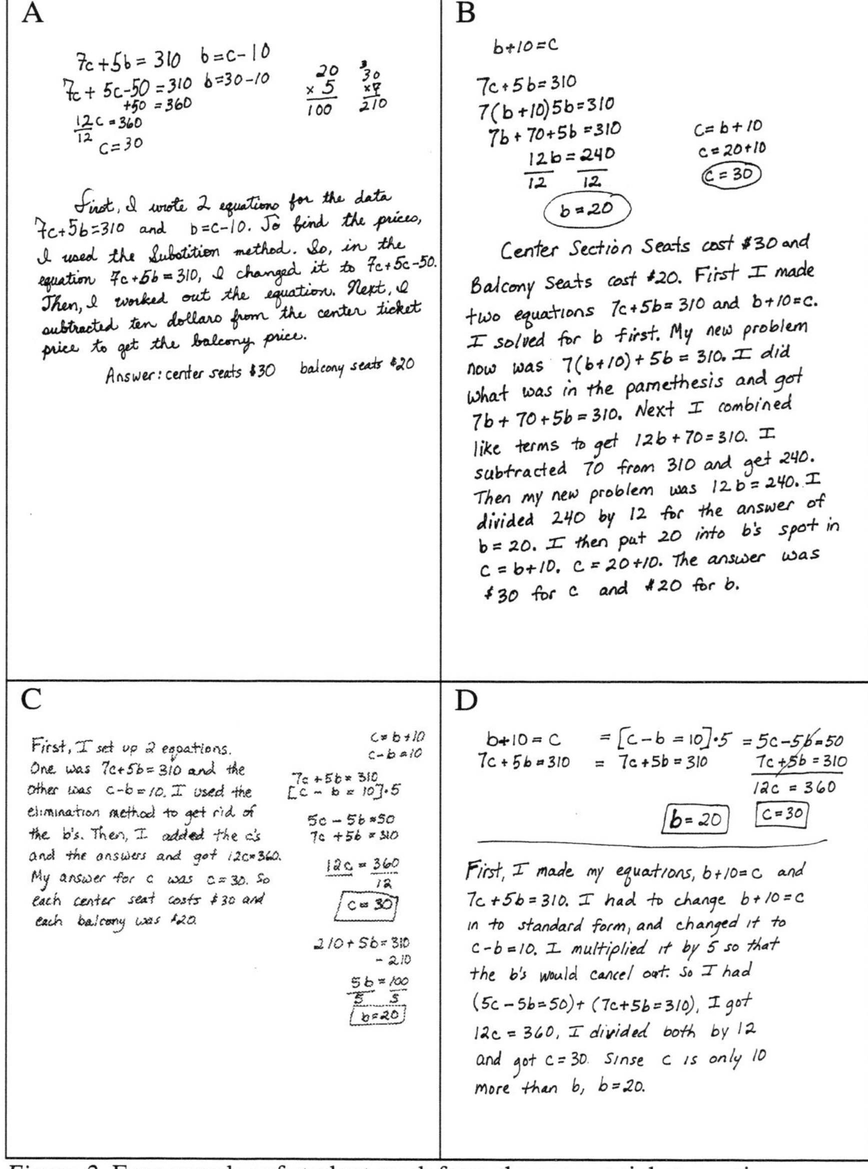

Figure 2. Four samples of student work from the concert ticket question.

As the Case Story process began, Charles distributed a complete set of student work samples to the members of his group without comment. The group members then had seven minutes to review the work in silence. During this time, members of the group recorded what they noticed and what they were wondering about. This work was followed by two 7-minute periods during which group members shared their noticings and wonderings respectively. During this time, Charles quietly listened and made note of the statements made by his peers. The noticings and wonderings shared by the group are shown in Figure 3.

NOTICINGS	WONDERINGS
a. I noticed that all the students used 1 of 2 methods (substitution or elimination). b. I noticed no one used a graphical method. c. I noticed no one made a table. d. I noticed all the students got the correct answer. e. I noticed students did not label or define their variables. f. I noticed explanations were a step-by-step of the procedure used to solve it. g. I noticed explanations did not explain why the model they created was appropriate or how the model connected to the context of the problem.	a. I'm wondering what the mathematical goals of the lesson were. b. I'm wondering what prior experiences students had before this lesson (i.e., which methods they had been taught and how they were taught them). c. I'm wondering whether this was an instructional task or an assessment task. d. I'm wondering how would the lesson and student work have been different if the task was used as an instructional or assessment task. e. I'm wondering what students understood about the models that they created. f. I'm wondering the extent to which students' explanations reflect instruction. g. I'm wondering what would happen if you gave this task at the beginning of a unit.

Figure 3. Noticings and wonderings shared during Charles' case story.

After group members shared their comments about the student work, the storyteller (Charles) had 10 minutes to share his perspective. At this point Charles could describe what he saw in the work, add other information that he felt was important to share, respond to some or all of the questions raised, or comment on anything surprising or unexpected that he heard. This phase of the Case Story was interactive and the group members were encouraged to engage in discussion with the storyteller.

Charles stated that he wanted to share his lesson goals with the group before addressing what they had "noticed" and "wondered" about the student work. Charles said his students had been solving simultaneous equations for several weeks and he was hoping the "Concert Tickets Question" would take them to the next level. Prior to this lesson his students had done few word problems, and none where the solution method (substitution, elimination, graphing) was not specified. In this problem, students had an opportunity to practice solving simultaneous equations in a meaningful context and to engage in a more "open ended" question, such as those on the state assessment.

Charles also talked with his peers about his desire to engage his students in a cognitively challenging mathematical task (see Smith & Stein, 1998). He explained that the "Concert Tickets Question," which did not prescribe a specific "pathway" for solving the problem, would force students to stop and think about what mathematics was needed. Additionally, by asking students to justify their solution method, he was hoping to learn about the thought process that his students used to create the mathematical model. Charles then addressed several of the "noticings" and "wonderings" raised by his peers.

The "noticings" and particularly the "wonderings" of the group (shown in Figure 3) raised questions about the instruction. Charles' responses to the issues raised indicated that he was reflecting on his own instruction—in particular, on the relationship between his instruction and his students' learning. For example, with respect to noticings "a" and "c" in Figure 3, Charles was not surprised that no one made a table, because students had learned to solve simultaneous equations in class using specific methods that had not included the use of tables. Therefore, although the task did not provide a specific solution "pathway," the students would naturally rely on the methods they had been learning in class.

As Charles responded to other issues raised by his colleagues, he continued to make connections between his own instruction, his students' work, and his students' mathematical thinking. For example, when reexamining students' explanations, Charles indicated that he was disappointed in what his students had written. He was looking for a written justification of the model – why it was chosen, what the terms meant, and what the sentences represented. Instead, he got what sounded like himself – standing at the board solving a system step-by-step (as noted in noticing "f" in Figure 3 and exemplified in the work shown in Figure 2). He did not get what he was hoping for in the explanations, which made him wonder if he had really asked for what he wanted, or if his students just did not know how to give him what he wanted because he had not modeled it. After listening to and reflecting on his peers' discussion, Charles indicated that he felt the students' explanations represented his instruction – a question that had been raised by one of his peers in wondering "f" in Figure 3. In previous lessons, he shared that he had not modeled the set-up of many 'real world' problems; when he used such problems in earlier units, he really did not engage students in this process. He just told them what he was doing so the class could get to the task of manipulating the equations to obtain the solutions. After hearing the group talk about the uniformity of the process-oriented explanations of his students (noticings "f" and "g" in Figure 3 and exemplified in the work shown in Figure 2), he began to see his instruction in his students' work. Charles concluded that he needed to give equal time to the model setup and the process in his lessons. One of the group members then mentioned that he could have the students work on only setting up models – leaving the process for some other time. Charles agreed that was a great idea.

Learning from Charles' Case Story

Several aspects of Charles' Case Story provide evidence that the Case Story protocol is a mechanism for teachers to: 1) engage in productive discussions about teaching and learning with colleagues; 2) inquire into their teaching practice; and 3) reflect on instruction and its effect on students' learning. For example, Charles and his peers engaged in a discussion that focused on what students were doing and thinking (e.g., noticings "a – g" and wondering "e" in Figure 3) and on the connection between students' learning and Charles' instruction (e.g., wonderings "b" and "f" in Figure 3).

The evidence suggests that, after the conversation with his colleagues, Charles was thinking differently about the lesson he had taught and the task he had used. When he came to the session that day, with this set of student work, he was feeling pretty good. He thought he had chosen a task that would make his students think mathematically and he thought their work showed what they could do. However, after listening to his group's discussion, he was beginning to think that he knew what his students could do, but he did not know how they thought – what they understood about using an algebraic system of equations to model and solve a real world problem.

Charles' Case Story also provides evidence that teachers were able to discuss teaching respectfully and to raise critical issues for Charles to consider. Over time, teachers became more attuned to what was important and how to raise issues in a way that was non-judgmental. Although it is difficult to determine what other teachers in the group might have learned from the 31 minutes spent focusing on Charles' instruction, the time provided another case for teachers to analyze, compare to their own teaching, and perhaps extract some general "lessons learned."

The Benefits of Using Case Stories with Teachers

Charles' Case Story was typical of the stories that emerged from the ESP teachers, in its focus on looking for evidence of students' mathematical thinking and linking teachers' instruction with their students' learning. Teachers repeatedly stated that Case Stories helped them focus their attention more directly on their students' thinking and how their instruction supported or inhibited the development of students' mathematical thinking. For example, one teacher claimed,

> I need to be asking myself about who is doing the talking and thinking when the students are trying to complete the task. I already know that many times I am the one thinking for them. This is something I really want to work on.

Furthermore, Case Stories offer teachers an opportunity to identify an area of their practice on which they could actively work to refine.

Finally, Case Stories provide a non-threatening way for teachers to talk about teaching and learning in their classrooms and to develop a critical stance towards teaching by reflecting on their own teaching and by listening to and respectfully questioning the practice of their colleagues. According to Ackerman and colleagues (1996), "...the Case Story process helps break down the isolation of practitioners and build a more collegial environment" (p. 23). This effect is exemplified in Charles' statement that he is "looking forward to sharing this month's assignment. It is valuable to share our lessons, successes, and struggles."

Using the Case Story Protocol also presents some challenges. Professional developers have practical issues to confront: How many teachers will come to a session prepared to discuss their practice? Can copies of materials that teachers need to distribute in order to tell their story be made at the last minute if they forgot to bring copies with them? Will the materials teachers bring engage teachers in productive conversations? One of the more substantive challenges is developing teachers' ability to focus their comments and reflections on important aspects of instruction and to identify issues non-judgmentally.

Towards this end, we initially placed a facilitator with each small group. The facilitator sometimes needed to use "I noticed" or "I'm wondering" comments to model and encourage constructive feedback and to cast critical comments (their own or from the group) in ways that would lead the storytellers to productive self-reflection. Over time teachers became quite skillful at respectfully identifying key issues for others to consider. As teachers became more experienced in this process, facilitators were not routinely placed with the teacher-groups.

Conclusion

The Case Story process was effective in encouraging teacher reflection and collaboration and supporting learning from practice. It is critical, however, to consider the Case Story process *as one component* of a coherent model for professional development intended to help teachers select and enact cognitively challenging mathematical tasks in their classrooms. Teachers came to the Case Story process with a set of frameworks and tools (e.g., the Mathematical Tasks Framework), which they had come to know and understand through the practice-based PD sessions in which they had engaged. These frameworks and tools provided a shared language for teachers to discuss teaching and learning and served to focus their reflection on aspects of instruction that matter most – what students are doing and thinking and how the actions and interactions between a teacher and her students shape their opportunities to learn.

We argue that the practice-based experiences in which teachers engaged during the PD sessions served as "a mediating device between teachers' reflection on their own practice and their ability to interpret their own practice as instances of more general patterns of task enactment" (Stein et al., 2000, p. 34). Through the analysis of narrative and video cases of teaching and student work during the PD sessions, teachers had the opportunity to resonate with what the featured teacher and students were doing and to make comparisons with their own practice. By closely examining the particulars of practice as represented in these artifacts of teaching, teachers began to develop powerful generalizations about teaching and learning (e.g., factors that support and inhibit student learning). The assignments that teachers were given following each session provided teachers with the opportunity to apply these generalizations to their own practice. The Case Story process, then, provided a mechanism for teachers to begin to discuss publicly their teaching and to continue to reflect on their practice with the help of like-minded colleagues who were able to raise issues for their fellow teachers to consider.

References

Ackerman, R., Maslin-Ostrowski, P., & Christensen, C. (1996). Case Stories: Telling tales about school. *Educational Leadership, 53*(6), 21-23.

Ball, D. L., & Cohen, D. K. (1999). Developing practice, developing practitioners: Towards a practice-based theory of professional education. In G. Sykes & L. Darling-Hammond (Eds.), *Teaching as the learning profession: Handbook of policy and practice* (pp. 3-32). San Francisco: Jossey-Bass.

Borasi, R., & Fonzi, J. (2002). Professional development that supports school mathematics reform. *Foundations monograph Volume 3.* Washington: The National Science Foundation.

Duncan, P. K., & Clayburn, C. (1997). *Why haven't I heard from you? Evoking the voices of adult learners through transformative teaching.* Paper presented at the Annual Meeting of the University Council for Educational Administration. Orlando, FL.

Hart, L. (1991). Assessing teacher change in the Atlanta Math Project. In R. G. Underhill (Ed.), *Proceedings for the thirteenth annual meeting of the North American Chapter of the International Group for the Psychology of Mathematics Education* (pp. 78-84). Blacksburg, VA: Virginia Tech.

Henningsen, M., & Stein, M. K. (1997). Mathematical tasks and student cognition: Classroom-based factors that support and inhibit high-level mathematical thinking. *Journal for Research in Mathematics Education, 28*(5), 524-549.

Maslin-Ostrowski, P., & Ackerman, R. H. (1998). Case story. In M. Galbraith (Ed.), *Adult learning methods: A guide for effective instruction.* Malabar, FL: Krieger Publishing.

National Council of Teachers of Mathematics. (2000). *Principles and standards for school mathematics*. Reston, VA: Author.

Schon, D. A. (1983). *The reflective practitioner*. New York: Basic Books.

Schon, D. A. (1987). *Educating the reflective practitioner*. San Francisco: Jossey-Bass Publishers.

Smith, M. S. (2000). Reflections on practice: Redefining success in mathematics teaching and learning. *Mathematics Teaching in the Middle School, 5*(6), 378-382; 386.

Smith, M. S. (2001). *Practice-based professional development for teachers of mathematics*. Reston, VA: National Council of Teachers of Mathematics.

Smith, M. S. (in press). Talking about teaching: A strategy for engaging teachers in conversations about their practice. In G. Zimmermann (Ed.), *Empowering mentors of teachers of mathematics*. Reston, VA: National Council of Teachers of Mathematics.

Smith, M. S., & Stein, M. K. (1998). Selecting and creating mathematical tasks: From research to practice. *Mathematics Teaching in the Middle School, 3*(5), 344-350.

Stein, M. K., Grover, B. W., & Henningsen, M. (1996). Building student capacity for mathematical thinking and reasoning: An analysis of mathematical tasks used in reform classrooms. *American Educational Research Journal, 33*, 455-488.

Stein, M. K., & Smith, M. S. (1998). Mathematical tasks as a framework for reflection: From research to practice. *Mathematics Teaching in the Middle School, 3*(4), 268-275.

Stein, M. K., Smith, M. S., Henningsen, M., & Silver, E. A. (2000). *Implementing standards-based mathematics instruction: A casebook for professional development*. New York: Teachers College Press.

Stigler, J., & Hiebert, J. (1999). *The teaching gap: Best ideas from the world's teachers for improving education in the classroom*. New York: The Free Press.

Stigler, J.W., & Hiebert, J. (2004). Improving mathematics teaching. *Educational Leadership, 61*(5), 12-16.

Wood, T., Cobb, P., & Yackel, E. (1991). Change in teaching mathematics: A case study. *American Educational Research Journal, 28*(3), 587-616

---

[i] This chapter was prepared under the sponsorship of the ESP Project, under a grant from the National Science Foundation (DUE 0301962) to Margaret S. Smith. Any opinions expressed herein are those of the authors and do not necessarily represent the views of the Foundation.

[ii] The MTF distinguishes three phases through which tasks pass as they unfold during a lesson (Stein et al., 1996): First, as they appear in curricular or instructional materials; next, as they are set up or announced by the teacher; and finally, as students actually go about working on the task. All of these, but especially the enactment phase, are viewed as important influences on what students actually learn. The MTF makes salient that the cognitive demands of a task can change as it passes from one phase to the next.

[iii] See Smith (in press) for a discussion of how the non-threatening phrases "I noticed" and "I'm wondering" can be used to raise issues related to teaching and learning.

Elizabeth K. Hughes is an Assistant Professor of Mathematics Education at the University of Northern Iowa. She is interested in designing practice-based learning experiences for teachers and examining the development of teachers' mathematical knowledge for teaching, in particular, teachers' capacity to attend to students' mathematical thinking in their planning and instructional practices.

Margaret S. Smith is a Professor of Mathematics Education at the University of Pittsburgh. Over the past decade she has been developing research-based materials for use in the professional development of mathematics teachers and studying what teachers learn from the professional development in which they engage.

Melissa Boston is an Assistant Professor of Mathematics Education in the School of Education at Duquesne University, where she teaches mathematics methods courses for preservice secondary mathematics teachers and mathematics content courses for preservice elementary teachers. Melissa's research focuses on teachers' learning and instructional change following their participation in professional development experiences.

Michael L. Hogel is Supervisor of Science and Technology Education at the Mt. Lebanon School District and a doctoral student in mathematics education at the University of Pittsburgh. He holds a bachelor's degree in chemical engineering from the University of Pittsburgh and a master's degree in education from Widener University.

Clark, K.
AMTE Monograph 5
Inquiry into Mathematics Teacher Education
©2008, pp. 85-95

8

Heeding the Call: History of Mathematics and the Preparation of Secondary Mathematics Teachers

Kathleen M. Clark
Florida State University

In this chapter I summarize research and policy recommendations that call for the inclusion of the history of mathematics in the preparation of preservice mathematics teachers (PSTs). Next, I share a model for the course "Using History in the Teaching of Mathematics," which was designed to encourage PSTs to study and consider including the history of mathematics in teaching. An example of one "topic exploration" from the course is presented. Lastly, I share sample student responses from the topic exploration for the purpose of illustrating PSTs' perspectives of their own mathematical knowledge and inclusion of history in their future teaching.

> One of the frustrating things about using history in your teaching is that if students would really tune in to what you are saying, they would understand things so much better. Instead, it is almost like we have trained them to sift through the contextual knowledge that we are giving and look for a formula so they can get the right answer.
>
> (John, journal entry, 4/12/07)

John, a preservice mathematics teacher (PST), experienced the frustration in the introductory quote while working with students in the classrooms in which he observed, tutored, and "guest taught" during the Spring 2007 semester. John noted that the students in his classes merely wanted the teacher to show them how to "get the right answer." In the same semester, John was simultaneously completing *Using History in the Teaching of Mathematics* (referred to as *Using History* for the remainder of this chapter). The aim of this course is to provide PSTs an initial experience with studying the history of mathematics as well as an experience with considering ways in which to incorporate history in their teaching. However, Leng (2006) observed that, "the decision to incorporate history into curriculum is [not] always a practical one....The integration of history into the mathematics curriculum must be executed circumspectly, and achieved with the expertise of highly competent educators" (p. 486). Considering and then incorporating the history of mathematics during the "guest teaching" requirement of their preservice practicum proved to be a challenge for the PSTs enrolled in *Using History*.

I have two primary goals for this chapter: (1) to summarize the research and policy recommendations that call for the inclusion of the history of mathematics in the preparation of PSTs; and (2) to describe a model of a course that prepares PSTs to study the history of mathematics for use in teaching. To share this model, I first describe the content for *Using History* at Florida State University based upon my experience during the 2006 – 2007 academic year. Then, I describe an example from the course content and sample student responses to illuminate the perspective of PSTs.

Research and Policy as Guiding Forces: The Call for Requiring History of Mathematics for Prospective Teachers

The standards identified by both the National Council of Teachers of Mathematics (NCTM) and the National Council for the Accreditation of Teacher Education (NCATE) are often considered as policy recommendations when designing mathematics teacher education programs. The *NCATE/NCTM Program Standards: Programs for Initial Preparation of Mathematics Teachers* (2003), at both the middle and secondary levels, describes content standards for seven mathematical strands. For programs to achieve "National Recognition" under the NCATE model, "the program report must demonstrate that at least 80% of all indicators are addressed and at least one indicator is addressed for each standard" (NCTM, 2007). The final indicator for each standard calls for teachers to "demonstrate knowledge of the historical development of [topics] including contributions from diverse cultures" (NCATE, p. 4). Thus, appropriate consideration of history of mathematics content knowledge is important to achieve national recognition. More important, however, is the argument that understanding and engaging in the study of the history of mathematics contributes to the mathematical and pedagogical preparation of mathematics teachers (e.g., Fauvel, 1991; Liu, 2003).

Even if mathematics education programs are not dependent upon NCATE program reports and reviews, such recommendations raise an important question: How does a mathematics teacher preparation program ensure either a course or multiple course experiences that provide access to the historical development of school mathematics topics? The *Mathematical Education of Teachers* (CBMS, 2001) argued that prospective teachers improving their knowledge of the history of mathematics is one way to expand their ability to "undertake, and then be able to challenge their students in ways that will lead them to reason and make sense of mathematics" (p. 99). The recommendations for including historical content in the preparation of both middle grades and high school mathematics teachers focus on providing the means for prospective teachers "to develop an eye for the ideas of mathematics that will be particularly challenging for their students" (p. 126). For example, historical perspectives about algebra and number theory can help teachers in the middle grades illuminate the development and acceptance of modern notation for number patterns and properties, algorithms for solutions of equations, fractions, negative numbers, irrational numbers, and complex numbers (CBMS, p. 126).

Using History in the Teaching of Mathematics: Attempts to Heed the Call

Professional organizations, such as NCTM, NCATE, the International Congress on Mathematical Education (ICME), the Mathematical Association of America (MAA), and the American Mathematical Society (AMS), are deeply committed to both the mathematical education of students and the mathematical preparation of their teachers. In addition to these groups, individuals have been interested for decades in "the changing views about what mathematical knowledge is needed to be an effective teacher" (CBMS, 2001, p. 3). In the previous section I reviewed how various groups and individuals consider ways in which the history of mathematics contributes to providing prospective teachers with topics "rich in mathematics and...deeper understanding of the scope and important processes of the subject they will teach" (CBMS, p. 128). In the sections that follow, I describe one version of a *Using History* course that was developed with policy and research recommendations in mind, as well as with a philosophy that calls for PSTs to engage with course content historically, mathematically, and pedagogically.

Context for the Course

In recent years, what constitutes a history of mathematics course has become the subject of discussion for different audiences focused on undergraduate mathematics teaching (Rickey, 2005). Given the general professional discussion taking place about the content of history of mathematics courses, the *Using History* course described in the remainder of this chapter provides a framework for others seeking to construct a similar experience for prospective mathematics teachers. *Using History* is required for all prospective middle grades and high school teachers in the mathematics education program at Florida State University. The course introduces prospective teachers to tools and resources that not only have the potential to enrich their own mathematical knowledge, but also provides experiences with a rich array of learning opportunities for their future students.

Foci of the Current Course

Since Fall 2006, *Using History* has been designed to encourage PSTs to think about the history of mathematics from three perspectives and to develop experience in identifying, creating, and using appropriate resources to integrate a historical perspective into their teaching of mathematics. Thus, throughout the course, PSTs are given opportunities to:

- work with the **mathematics** that evolved over time, with respect to middle grades and high school mathematics content found in school curricula today;

- study and discuss the **historical and cultural** influences on and resulting from the mathematics being developed; and
- develop the **pedagogical** knowledge needed to integrate a historical perspective in the teaching of school mathematics.

In order to provide PSTs with meaningful experiences that address the course foci, it is essential to select mathematical topics and primary course tasks carefully.

Selection of Mathematical Topics

The choice of topics in a course like *Using History* is critical for several reasons. First, the topics must be significantly related to the mathematics prospective teachers will be expected to teach. If the content diverges too greatly from what PSTs perceive as significant to the content of their future teaching, they may disengage from considering the use of a historical perspective. Topic choices become more complicated when prospective middle and high school teachers are enrolled in the same course, as is the case at Florida State University. The complication primarily results from a discrepancy between what PSTs perceive as necessary content for teaching and what experts suggest as necessary (CBMS, 2001). For example, many prospective middle grades teachers do not envision themselves teaching a course in geometry. Therefore, focusing on various proofs of the Pythagorean Theorem is beyond the scope of what they feel is necessary during their teacher preparation program. In this way – even in a history of mathematics course – the PST becomes the arbiter of what constitutes appropriate mathematics preparation for teaching. The mathematical content of *Using History* provides a delicate balance of relevance to both certification levels and includes course tasks that enable PSTs to "develop a thorough mastery of the mathematics in several grades beyond that which they expect to teach" (CBMS, 2001, p. 7).

The second reason that topic selection is crucial relates to the philosophy from which the course has been designed. Indeed, *Using History* was created from the perspective that PSTs would study the humanistic aspects of the development of mathematics *and* engage in "doing" mathematics in historically significant ways. Consequently, the topics examined in the course need to possess historically-rich backgrounds, contribute to PSTs' mathematical preparation, and align with curricular expectations.

Finally, many mathematical topics that secondary students find difficult to understand conceptually possess a historical development that mirrors those difficulties (Radford, 2000). Many PSTs may not recall their own experience with these conceptual difficulties. Therefore, addressing such topics in a *Using History* course is one way to create an awareness of these difficulties. Sfard (1994) noted that obstacles for students frequently arise at those critical junctures where teachers who have already passed the juncture and mastered an idea cannot communicate with students who have not. She observed that "there is no better way to analyze this problem than by scrutinizing the historical development of mathematical knowledge. History is the best instrument for detecting invisible conceptual pitfalls" (p. 129).

Table 1 provides the order of mathematical topics from the most recent *Using History* course.

Table 1: *Using History - Course Topics and Length of Time Spent*

Topic	Time spent (number of 75-minute class sessions)
Use of history in the secondary classroom: What do the experts say?	4
Representation of quantities	3
Zero	1
Negative numbers	2
Solving linear equations	2
Solving quadratic equations	1
Solving cubic equations and complex numbers	2
The Pythagorean Theorem and Euclid's *Elements*	2
Development of trigonometry and trigonometric identities	2
Pascal's *Arithmetical Triangle* and probability	2
Concept of infinity	1
Introduction to the derivative	1
TOTAL	23

Course Philosophy and Significant Tasks

In the opening paragraph of the course syllabus, I identify the purpose of the course and within that statement, I reveal the philosophy from which I view the course:

> David Pimm (1983) once observed that, "the beauty of the study of the history of mathematics is that it can give a sense of place...from which to learn mathematics, rather than merely acquiring a set of disembodied concepts" (p. 14). The goal of this course is to engage each of you in the study of the history of topics that you will be expected to teach.... In addition, I hope that each of you will gain expertise in identifying and creating appropriate resources for the

purpose of integrating a historical perspective in teaching mathematics.

To provide PSTs with such an experience, I use five primary course tasks described in Table 2.

Table 2: *Using History - Primary Course Tasks*

Task	Brief description	Foci addressed
Key Topic Explorations	Course readings, supplemented with task(s) focusing on the topic with a historical emphasis (weekly)	Mathematical Historical/cultural Pedagogical
Journal Assignment	PSTs record their mathematical understandings, new knowledge, and pedagogical plans for incorporating the history of mathematics in future teaching (weekly)	Mathematical Historical/cultural Pedagogical
Library Assignments	Develop research skills for obtaining knowledge for the various course assignments; build skills for considering the use of the history of mathematics in teaching (two key tasks)	Mathematical Historical/cultural (with pedagogical influences)
"Book Club" Experience	Reading of a pre-selected text focused on the development of a specific mathematical idea; PSTs plan for and conduct discussions on the text and write reflections on experiences with the text (~6 weeks)	Mathematical Historical/cultural Pedagogical
Model Lesson Assignment	Culminating course task; written components include: historical background of the mathematical topic chosen; lesson plan and all accompanying documentation; and a bibliography of all resources used	Mathematical Historical/cultural Pedagogical

Sample Topic Exploration: Perceptions and Knowledge of PSTs

In this section I describe an example of a Key Topic Exploration through selected PSTs' reflections and work related to the task. Solving cubic equations is often considered one of the critical junctures in the development of mathematics that may be suitably addressed from a historical perspective (V. Katz, personal communication, 2/02/07). The solution of the cubic is a secondary mathematics topic that is typically taught devoid of its rich historical development. And, as evidenced by a content diagnostic administered to *Using History* students, even those preparing to teach secondary mathematics lack sufficient mastery of solving higher-order polynomial equations. Of 28 PSTs taking a content diagnostic in Spring 2007, only one successfully solved the equation, $x^3 + 4x^2 - 5x - 8 = 0$. Thus, in addition to experts identifying solving cubic equations as posing a critical juncture in secondary mathematics learning, the topic also represented a gap in the mathematical preparation of the PSTs enrolled in *Using History*.

To investigate solving cubic equations, PSTs were first asked to read an account of the historical development of solving the cubic equation (Berlinghoff & Gouvêa, 2004, p. 133-136). In addition, PSTs were given an excerpt from the Polynomials Module of the *Historical Modules for the Teaching and Learning of Mathematics* (Agwu, Frey, Greer, Taylor, Gould, Perkins, et al., 2005). This excerpt contains additional detail for the method of solution proposed by Girolamo Cardano in 1545, including a step-by-step solution of a cubic equation showing each numerical manipulation from the application of Cardano's formula.

Cardano's method for solving the cubic equation $x^3 = cx + d$ (as it appears in the Polynomials Module supplement) is given by:

$$x = \sqrt[3]{\frac{d}{2} + \sqrt{\left(\frac{d}{2}\right)^2 - \left(\frac{c}{3}\right)^3}} + \sqrt[3]{\frac{d}{2} - \sqrt{\left(\frac{d}{2}\right)^2 - \left(\frac{c}{3}\right)^3}},$$

and is in a slightly different form than that given in the course text. Consequently, part of the task for this particular topic exploration asked students: (1) to determine that the formula in their textbook

$$x = \sqrt[3]{-\frac{q}{2} + \sqrt{\frac{q^2}{4} + \frac{p^3}{27}}} + \sqrt[3]{-\frac{q}{2} - \sqrt{\frac{q^2}{4} + \frac{p^3}{27}}}, \text{ when } x^3 + px + q = 0,$$

is equivalent to the formula given in the Polynomials Module Supplement; and (2) to use either formula to solve the equation $x^3 + 9x - 26 = 0$. In addition, the task required PSTs to:

> Comment on the potential impact of integrating the history of solving cubic equations – which also contributed to the development of complex numbers – on student engagement and understanding when learning to solve cubic equations in secondary school mathematics.

Several of the PSTs' attempts to integrate the historical development of the solution to the cubic reflected their own difficulty or lack of experience with this topic. Janine, preparing to teach high school mathematics, observed:

> Incorporating the history of cubic equations would be beneficial to students because by using this equation I was able to solve a cubic equation for the first time in my life. If students learned the history,

> they would be able to see that solving these equations is possible, which may be motivation enough for some students to learn. (Cubic Equations Task, 3/13/07)

Similarly, Diane's prior experience with solving cubic equations was limited in other ways:

> This [Cardano's] formula is definitely more time consuming and difficult than using a method such as substitution. With Cardano's formula you aren't even sure about the numbers without a calculator. Then you have to deal with decimals, wherein substitution you have (usually) only whole numbers or fractions. If you could get actual numbers from the equation [Cardano's] it would then be more useful than Descartes Rule of Signs or the Rational Root Theorem. (Cubic Equations Task, 3/13/07)

Engaging with the history of solving cubic equations revealed gaps in both Janine's and Diane's mathematical preparation for teaching. In Janine's case, she was explicitly aware of not being able to solve a problem that her students may one day be required to solve. In the case of Diane, however, her response revealed significant misunderstandings regarding the solution of polynomial equations in general. The inclusion of tasks such as the Cubic Equations Task Exploration was beneficial for me as the instructor of this course; it enabled me to identify my PSTs' difficulties with secondary mathematics topics and to design instruction to focus on their needs.

This particular Key Topic Exploration also dominated PSTs' journal entries after the topic was investigated during class. Of the 20 (out of 28) PSTs who dedicated at least some portion of a journal entry to their experience with solving cubic equations, ten focused extensively on their own learning or ability to present this topic historically in teaching. Kristina admitted, "This is embarrassing, but to be honest, I really do not remember learning about cubic equations and how to solve them" (Journal Entry, 3/15/07). Similarly, Donna observed:

> During our diagnostic test [in] the first class session, I was completely lost when we were given a cubic equation. I found that learning this historical method of solving cubic equations gave me a greater understanding of how to solve cubic equations. My teachers in high school seemed to ignore this crucial area in mathematics and more and more I find that to be a detriment in my understanding of the subject. (Journal Entry, 3/18/07)

Finally, Daniel shared a struggle between his positive experience with a difficult topic and his beliefs related to appropriate uses of history in teaching:

> Working this week with the cubic equations gave me some insight into what degree of understanding I have on the subject. I think that when we know this and when we have an understanding of how the solution came to be that we must be wise in knowing when to apply it. If I were in a classroom where I felt the students could understand the material I would not hesitate in implementing this use of history. (Journal Entry, 3/15/07)

An examination of the journal excerpts provides insight into what PSTs perceive about their own learning of topics that they will eventually teach. The excerpts also include their perceptions of the impact of the use of history on potential student learning. Ten PSTs articulated more extensive mathematical, historical, and pedagogical treatments of solving cubic equations; six others chose to focus on more humanistic aspects of the historical development leading to the solution of the cubic equation. For these students, the "story" behind this advanced mathematical topic contributed to their engagement and enabled them to consider how their future students could benefit:

> I found the timeline of cubic equations a very dramatic scene. I guess though, in those times, and even still today, it is a big deal if one discovers something. When teaching a unit on cubic equations, I think telling this story, or having the students create a short story about this adventure, would engage them and they would want to learn the material more. If nothing else, hopefully they would gain an understanding as to why cubic equations are important. (Janine, Journal Entry, 3/15/07)

Janine's supposition that students would "hopefully…gain an understanding" is also an aspect of using the history of mathematics that is well-documented in the work of those supporting its use in teaching (e.g., Lingard, 2000, p. 41).

Conclusion

Upon reflection about the key content and tasks in *Using History*, I identified two dominant challenges, each of which intimately relate to confronting beliefs about teaching. First, I found it essential that the course enable PSTs to realize that important mathematical understanding can result from a historical treatment of key concepts. Almost weekly during the *Using History* course, more than one PST highlighted issues they believed would prevent the application of a historical perspective in teaching. For example, many PSTs expressed beliefs that student ability (or lack thereof) combined with a perceived level of difficulty of learning through a historical perspective would impact their decision to use the history of mathematics in their teaching. If prospective teachers can identify how their own understanding is enriched by studying a topic historically, they may begin to consider alternative perspectives in teaching. Consequently, I continue to develop Key Topic Explorations that include content found in the benchmarks of the Florida Next Generation Sunshine State Standards for K-12 mathematics *and* that provide opportunities for PSTs to identify a positive impact on their own learning that is rooted in the historical treatment of mathematical topics.

Second, I believe many PSTs saw the topic explorations and aspects of the major course tasks as activities that could simply be transported to a middle or high school classroom. This perception influenced their views about the "usefulness" of the history of mathematics in teaching. For example, if a PST considers a task too challenging, then (s)he may consider that aspect of history not appropriate for secondary students. Instead, I intended for the *Using History* students to engage in the mathematical and instructional ideas related to a given topic. Then, the larger task was for each PST to consider what content and tasks could be translated into future practice. Thus, a more explicit effort in future iterations of the course will be to move PSTs from the stance of direct transmission of what they do in *Using History* to one in which they view the course as a resource to inform their future instruction.

References

Agwu, N., Frey, P., Greer, T., Taylor, G., Gould, C., Perkins. J., et al. (2005). Polynomials. In V. J. Katz & K. D. Michalowicz (Eds.), *Historical modules for the teaching and learning of mathematics.* Washington, DC: The Mathematical Association of America.

Berlinghoff, W. P., & Gouvêa, F. Q. (2004). *Math through the ages: A gentle history for teachers and others.* Farmington, ME/Washington, DC: Oxton House/The Mathematical Association of America.

Conference Board of the Mathematical Sciences (CBMS). (2001). *The mathematical education of teachers: Vol. 11. Issues in mathematics education.* Providence, RI: American Mathematical Society.

Fauvel, J. (1991). Using history in mathematics education. *For the Learning of Mathematics, 11*(2), 3-6.

Leng, N. W. (2006). Effects of an ancient Chinese mathematics enrichment programme on secondary school students' achievement in mathematics. *International Journal of Science and Mathematics Education, 4*(2), 485-511.

Lingard, D. (2000). The history of mathematics: An essential component of mathematics curriculum at all levels. *Australian Mathematics Teacher, 56(1),* 40-44.

Liu, P.-H. (2003). Do teachers need to incorporate the history of mathematics in their teaching? *Mathematics Teacher, 96*(6), 416-421.

National Council for the Accreditation of Teacher Education (2003). *NCATE/NCTM Program standards: Programs for initial preparation of mathematics teachers.* Washington, DC: National Council for the Accreditation of Teacher Education.

National Council of Teachers of Mathematics (n.d.). *NCATE mathematics program standards.* Retrieved May 2, 2007, from http://www.nctm.org/standards/content.aspx?id=2978&ekmensel=c580fa7b_10_0_2978_3

Pimm, D. (1983). Why the history and philosophy of mathematics should not be rated X. *For the Learning of Mathematics, 3*(1), 12-15.

Radford, L. (2000). Historical formation and student understanding of mathematics. In J. Fauvel & J. van Maanen (Eds.), *History in mathematics education: The ICMI study* (pp. 143-170). Dordrecht, The Netherlands: Kluwer Academic.

Rickey, V. F. (2005). *Teaching a course in the history of mathematics.* Retrieved April 30, 2007, from http://www.math.usma.edu/people/rickey/hm/mini/default.html

Sfard, A. (1994). What history of mathematics has to offer the psychology of mathematical thinking. In J. P. da Ponte & J. F. Matos (Eds.), *Proceedings of the 18th International Group for Psychology in Mathematics Education* (Vol. I, pp. 129-132). Lisbon, Portugal: Program Committee of the 18th PME Conference.

Kathleen M. Clark is an Assistant Professor of Mathematics Education in the School of Teacher Education at Florida State University. Her work includes teaching and mentoring mathematics education students, serving as a core faculty member in the FSU-Teach Program, and conducting inquiry about the inclusion of history in teaching mathematics.

Hjalmarson, M., and Suh, J.
AMTE Monograph 5
Inquiry into Mathematics Teacher Education
©2008, pp. 97-107

9

Developing Mathematical Pedagogical Knowledge by Evaluating Instructional Materials

Margret A. Hjalmarson
Jennifer M. Suh
George Mason University

In this chapter, we describe how artifacts of teaching practice can be used to elicit preservice teachers' understanding of mathematics and mathematics teaching. We gave two assignments to preservice teachers from two mathematics methods courses (elementary and secondary). The elementary assignment asked the preservice teachers to evaluate virtual manipulatives. The secondary assignment asked the preservice teachers to compare different types of curricula. Both assignments asked preservice teachers to explain how the materials could be useful for teaching practice.

A wealth of resources is available for teachers to use and examine online and in print. As a result, developing teachers' ability to evaluate materials is an important aspect of their mathematical pedagogical knowledge for teaching. The analysis of materials combines an understanding of mathematics content, the nature of learning, and the nature of teaching. In this chapter, we present two studies of preservice teachers' analysis of materials for teaching (e.g., curriculum, technology tools). We independently designed tasks for mathematics methods courses we taught at the elementary and secondary levels, and then found that there were shared characteristics of the tasks independent of the course. Analysis of student responses indicated common characteristics of mathematics teacher learning independent of grade level. The purpose of this chapter is to describe a design process for eliciting preservice teachers' mathematical understanding in a pedagogical context.

Research indicates that teacher education programs should include practice-based activities to help teachers learn how to teach mathematics effectively (Ball & Cohen, 1999; Darling-Hammond, 1998; Lampert & Ball, 1998; Wilson & Berne, 1999). Franke and Chan (2006) argue that we need to choose "high-leverage practices," or those aspects of mathematics teaching practice that are central to supporting the development of mathematical understanding, as productive starting places for novice teachers. One of the high-leverage skills for mathematical knowledge for teaching is to make judgments about the mathematical quality of instructional materials and modify them as necessary

(Ball, 2003). Lloyd and Behm (2005) found that preservice teachers' critical analysis of instructional materials provided opportunities to develop mathematical pedagogical knowledge and familiarity with the curriculum. Shulman (1987) described pedagogical content knowledge as the ability to communicate the "most useful forms of representation of these ideas, the most powerful analogies, illustrations, examples, explanations, and demonstrations – in a word, the ways of representing and formulating the subject that make it comprehensible to others" (p. 9). Yet, novice teachers have difficulty communicating mathematical concepts for teaching because they either lack the mathematics content knowledge or the pedagogical knowledge. By evaluating and selecting best instructional materials or models, preservice teachers can develop this mathematical pedagogical knowledge.

Curricular knowledge includes teachers' understanding of the nature of curriculum and the purposes and functionality of materials for teaching (Shulman, 1986). Teachers' use of curriculum is driven by a variety of factors, including their knowledge, beliefs, and experience (Remillard, 1999). Research related to reform-oriented instructional materials and technology integration indicates that preservice teachers are often challenged because many of them never experienced learning or teaching in that way (Ball & Cohen, 1996; Lloyd, 2002; Remillard, 2000). Lloyd and Behm's (2005) study of preservice elementary teachers' learning with middle-school curriculum materials in undergraduate mathematics courses found that the curriculum materials were different from the mathematics textbooks they previously used. In terms of technology, Battey, Kafai, and Franke (2005) studied preservice teachers' criteria for evaluating and using mathematical software. They found that most preservice teachers focused on surface features, such as clear directions, rather than focusing on the content or pedagogical issues. Statements made by the preservice teachers indicated a concern for general learning, engagement, and motivation, rather than specific mathematical content.

Providing preservice teachers experience with discriminating and evaluating instructional materials can help them become more confident and competent to teach mathematics. This study focused on two tasks designed to provide preservice teachers with opportunities to analyze instructional materials (textbooks and mathematical applets) to develop this high-leverage, practice-based skill. We first describe the students' responses to a textbook analysis task conducted in a secondary mathematics methods course. Next, we discuss a mathematical applet task used in an elementary mathematics course to elicit preservice teachers' understanding of different representations of mathematical content and how to incorporate them into instruction.

Textbook Analysis

The textbook analysis task was designed for a secondary mathematics methods course. Because textbooks play an important role in secondary teaching and because preservice teachers were unfamiliar with available texts, the assignment was designed both to expose them to different textbooks and to elicit their beliefs about learning mathematics. Traditional textbooks include an explanation of a topic followed by practice problems. Reform textbooks (e.g., *NSF*-supported, *Standards*-based) use an activity-based or problem-based approach for K-12 students to develop understanding. The preservice teachers analyzed units on the same content area from each type of textbook and described the general character of the texts. The preservice teachers responded to the following questions for the analysis: 1) Select a mathematical topic (e.g., addition of fractions) and compare the textbooks' treatment of the topic. How are they similar? How are they different?; 2) How would you, as a teacher, structure your teaching differently with each type of textbook?; and 3) How do the textbooks represent mathematics differently? How do the textbooks represent mathematical problem solving differently?

Twenty-two preservice teachers completed the textbook analysis assignment across two semesters (14 men, 8 women). Nine participants were employed as mathematics teachers (five in public high school, one in private high school, three in middle school). Because these nine were practicing teachers, some of their comments related to a textbook they were currently using to teach mathematics.

Textbook Analysis Results

The participants' beliefs about mathematics teaching fell into three categories: real-world applications, skills and algorithms, and a blend of the aforementioned approaches (Hjalmarson, 2005). Some preservice teachers recognized the importance of real world applications in mathematics learning as a venue for making mathematics meaningful and purposeful to their students. The *Standards*-based textbooks often fill the need for units based in real world applications. The second category of responses were from advocates of skill-based or algorithmic approaches because students need to understand the steps and procedures of mathematics. A third group envisioned a blended teaching approach between applications and skills. The third approach was an attempt to mediate the need for conceptual and procedural understanding.

Classroom discussion about the textbook analysis helped teachers see diverse perspectives about curriculum. Part of the in-class discussion focused on the purposes for textbooks and the nature of mathematics learning in each context, and was reflected in the teachers' responses:

> The [reform-based] book focuses on the concepts and how they relate to other mathematical concepts. The primary focus is not on regurgitating a formula to find answers. I believe that the [traditional textbook] has a tendency to promote just finding the answers.

Reform-based textbooks are often designed around problems and real-world contexts and the materials themselves may not contain formulas or definitions because the problems are intended to help students develop such formulas and definitions. However, some of the preservice teachers had used textbooks as reference guides to help them solve problems. Reform-based textbooks are not intended to be reference guides in the sense that a dictionary is a reference. Given their own experiences in using a textbook to learn mathematics, we want our preservice teachers to grapple with two main questions: 1) From a learning perspective, how is instruction balanced between concepts explained in a textbook and concepts explored via inquiry and investigation in real-world contexts?; and 2) How do different types of textbooks support different modes of instruction?

Mathematics Applets Analysis

The mathematics applets analysis was designed for elementary preservice teachers taking an elementary mathematics methods class. The 22 preservice teachers, 20 women and 2 men, were placed in local elementary schools as interns. The mathematics applet analysis involved three processes: 1) evaluating effective representations and affordances for mathematical thinking; 2) evaluating and selecting the appropriate model for teaching using technology; and 3) collaborative lesson planning using technology. We expected these processes to help teachers develop the ability to discriminate among multiple models and make judgments on the clarity, effectiveness, and appropriateness of the representations.

For Process 1, the instructor engaged the preservice teachers in a discussion about selecting effective representations by examining specific criteria. She posed two questions related to the chart in Figure 1: 1) Does the representation have transparency, efficiency, generality, clarity, and precision? (see National Research Council, 2001, pp. 99-101 for definition of terms); and 2) What mathematical thinking opportunities are afforded through the mathematics applets (e.g., understanding, applying, problem solving and reasoning, making and testing conjectures, and/or creating)?

As a homework assignment (Process 2), preservice teachers found at least four mathematics applets that focused on one concept. They were asked to consider the following questions taken from *Young Mathematicians at Work* (Cameron, Dolk, Fosnot, Hersch, & Werner, 2006) as they compared and analyzed the applets.

1. What are the different uses of these mathematical models?
2. How are students' developmental needs supported by the different uses of these models?
3. Is there a developmental progression for these models? Are some uses of the model precursors to others?
4. What role does the model play in helping students visualize different strategies?
5. How might one represent a given strategy with each model and mathematical applet? (p. 27)

<table>
<tr><td>Website</td><td colspan="3"></td></tr>
<tr><td>Applet name</td><td></td><td>Grade level</td><td></td></tr>
<tr><td>Description of mathematical concept</td><td colspan="3"></td></tr>
<tr><td colspan="4">__Concept tutorial/Skill Practice __ Investigation/problem solving
__Open exploration</td></tr>
<tr><td colspan="4">Analysis of Mathematical Representations and Models
(5 ***** excellent – 1*poor)</td></tr>
<tr><td colspan="3">Transparency:
How easily can the idea be seen through the representation?</td><td></td></tr>
<tr><td colspan="3">Efficiency:
Does the representation support efficient communication and use?</td><td></td></tr>
<tr><td colspan="3">Generality:
Does the representation apply to broad classes of objects or concepts?</td><td></td></tr>
<tr><td colspan="3">Clarity:
Is the representation unambiguous and easy to use?</td><td></td></tr>
<tr><td colspan="3">Precision:
How close is the representation to the exact value?</td><td></td></tr>
<tr><td colspan="4">Mathematical thinking opportunities afforded by the mathematics applet</td></tr>
<tr><td colspan="2">Connecting: Constructing conceptual connections and multiple representations</td><td colspan="2"></td></tr>
<tr><td colspan="2">Applying: Carrying out or using procedural knowledge flexibly & efficiently</td><td colspan="2"></td></tr>
<tr><td colspan="2">Problem Solving and reasoning: Formulating, representing and solving mathematical problems</td><td colspan="2"></td></tr>
<tr><td colspan="2">Making and testing conjectures: Making conjectures or judgments about mathematical ideas</td><td colspan="2"></td></tr>
<tr><td colspan="2">Creating: Putting elements together to generate, plan, or produce mathematical ideas</td><td colspan="2"></td></tr>
</table>

Figure 1. Checklist for evaluating mathematics applets.

In the final phase (Process 3), the preservice teachers worked in groups planning a lesson for a specific concept (e.g., subtraction with regrouping). The lesson planning allowed the preservice teachers to discuss the salient aspects of different mathematics applets as well as their selection criteria and analysis, using the previous five questions to plan an effective lesson.

Mathematics Applet Analysis Results

The preservice teachers found a variety of mathematical models and representations available on the web and identified the uses of these models using their methods textbook. They critically examined each model using the effective representations criteria and determined the developmental needs of the learners when using these models. In doing so, they referred to the state and national standards to familiarize themselves with the objectives at each grade level.

Preservice teachers' analysis of models using technology showed the use of their mathematical knowledge for teaching. The major themes that emerged were a) the ability to identify, represent, and explain a concept with multiple models, and b) an understanding of the developmental levels for different mathematical models. They commented that the mathematics applets helped them visualize abstract mathematical ideas with interactive and dynamic pictorial models and new, nontraditional ways to model math ideas. By considering the prerequisite knowledge necessary for students to engage with each applet, the preservice teachers had to consider students' developmental levels. This consideration helped the preservice teachers determine which mathematics applet would be appropriate for different grade levels and led to important discussions about differentiation and tiered learning for different learning styles. They could readily see how these materials could be used for visual learners, students with special needs, and English Language Learners, all of whom may need the support of pictorial representations.

For example, one teacher described in her written analysis how the different technology applets represented four different models of the base ten system: proportional vs. non-proportional models; number line (measurement model) vs. chip trading model (composing and decomposing numbers). In addition, this teacher commented on how the proportional model, *Base Block Addition* (National Library of Virtual Manipulatives, 2008a), is the easiest type of model to use when thinking about values (see Figure 2). However, non-proportional models are also essential to understanding the complexity of our number system. For example, money applets (Arcytech, 2003) can be used to assist students in learning to compute with money. In considering which model to use when teaching skip-counting, the preservice teacher felt that number line manipulatives (e.g., Freudenthal Institute, 2008) were easiest to use. For modeling composing and decomposing numbers, Base block addition and Chip Abacus (National Library of Virtual Manipulatives, 2008a, 2008b) illustrated the concept explicitly.

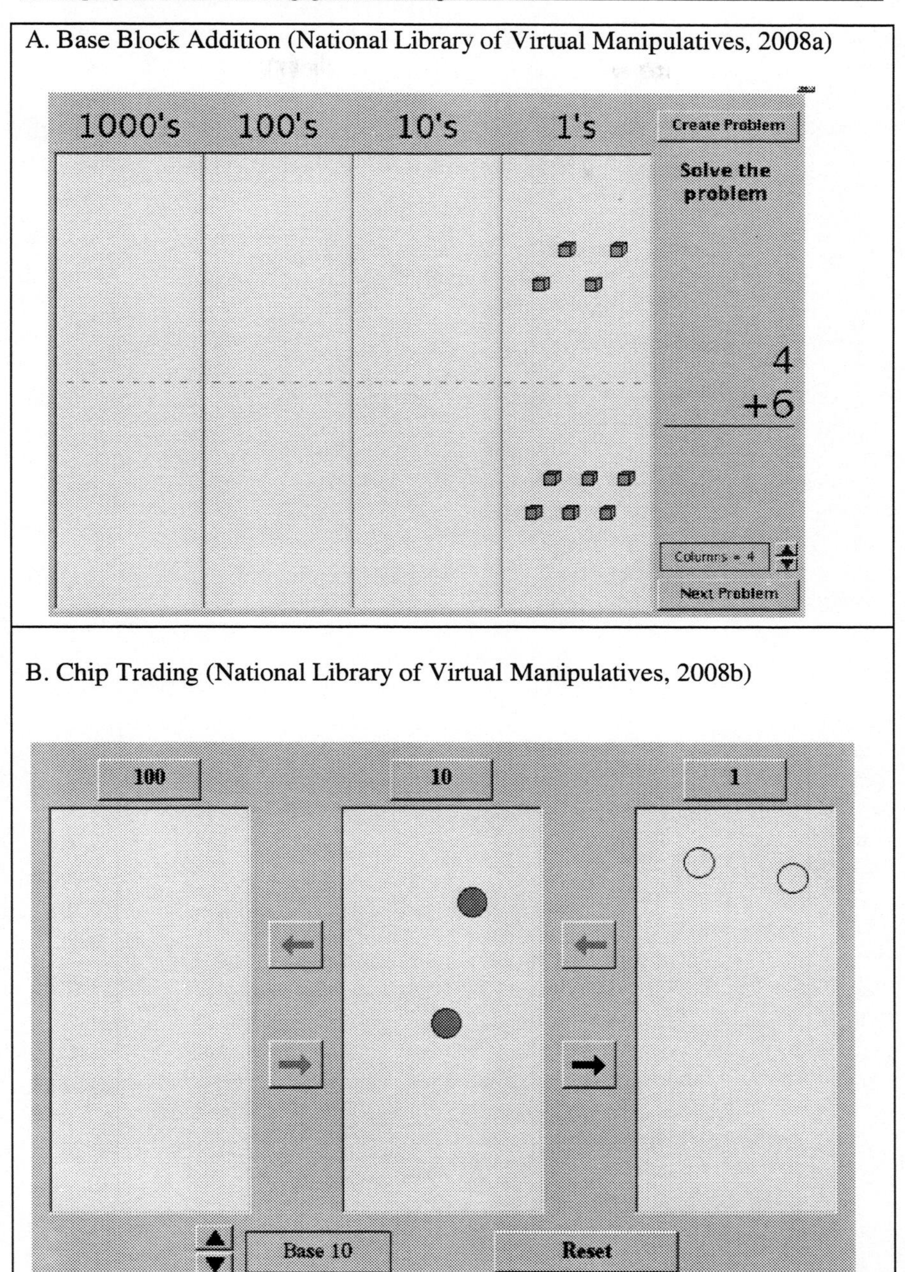

Figure 2. Examples of virtual manipulatives.

Through this task, the preservice teachers recognized that there are various models for illustrating a mathematical concept. In addition, they began to understand that different models supported different developmental progressions for students. Furthermore, by considering how to represent a given strategy with each model in a lesson, preservice teachers applied mathematical knowledge for teaching.

Common Design Characteristics for Promoting Pedagogical Content Knowledge

The central characteristic for both activities required that the preservice teachers think about mathematics from a pedagogical perspective. Although one task used secondary textbooks and the other used technological mathematical models, both elicited preservice teachers' beliefs about mathematics as well as the development of curricular knowledge. The textbook analysis task elicited preservice teachers' beliefs about mathematics content and the framing of that content for students. For the mathematics applets task, the preservice teachers used knowledge of mathematics and student development to evaluate representations of mathematical models. In both tasks, we asked teachers to consider mathematics pedagogically and pedagogy mathematically.

The design considerations for the tasks could be included in other tasks for preservice teachers. First, we placed a premium on experiences for the preservice teachers that would be practice-based – the task should represent a typical aspect of a teacher's job. In many mathematics methods courses, a major focus is often on lesson planning. However, the work of teachers goes beyond lesson planning and includes not only the creation of materials but also the use of existing materials. Curricular knowledge includes the analysis, adaptation, and understanding of the characteristics of instructional materials (Remillard, 1999; Shulman, 1986).

Second, both tasks required that the preservice teachers discriminate between materials by comparing them and describing their potential use in the classroom. We focused on readily-available materials to increase the meaningfulness of the task. The evaluation of materials, particularly in light of a growing variety of both paper-based and computer-based materials, is a task teachers at all levels undertake. Curricular considerations should include analyses of influence on student learning, the nature of mathematics represented in the materials, and how the materials integrate with other products. The evaluation process for both types of materials we described here elicited the preservice teachers' beliefs about the role of such materials in the classroom and their vision for mathematics teaching.

Third, these tasks supported the preservice teachers as they engaged in discourse about instructional materials, a necessary skill for effective teaching practice. Secondary teachers collaborate within a department or with a group of teachers teaching the same course. Elementary teachers collaborate with other teachers in their grade level, mathematics specialists, special education teachers, and ELL specialists. As a result of these assignments, the preservice teachers discussed their individual analyses of the materials in whole-class discussion. This process allowed them to share and to defend the reasoning they use in evaluating curricular materials. This process included both revealing their understanding of the purposes for the curricular materials as well as how they would implement them with students.

Finally, these tasks required the preservice teachers to identify the mathematics in the textbooks and the applets. This skill is critical for developing mathematics knowledge for teaching and analyzing how different materials represent fundamental aspects of mathematics in different ways. The mathematical analyses moved the discussion beyond the aesthetic, surface features of the materials toward student learning and instruction.

Conclusion

These analyses of curricular materials provided our preservice teachers with tangible tools for teaching. Due to the wealth of on-line materials and the regular changes in state and local curriculum guides, the analysis of curricular materials is a part of every teacher's job. Other items that preservice teachers could analyze for similar assignments include state and local standards documents and curriculum frameworks, lesson planning resources, and assessment tools. Commonalities found in the two tasks we described in this chapter indicate that both the textbook analysis and the mathematics applets analysis are practice-based tasks that can be implemented in both elementary and secondary preservice methods courses to promote mathematical pedagogical knowledge.

Knowledge of curriculum and curriculum materials, as well as effective integration of technology in learning mathematics, is necessary for teachers across the grade bands. Teachers at all levels must be able to make judgments about the mathematical quality of instructional materials and represent ideas carefully using multiple models, which are often technologically-based, in order to maximize student learning. Through explicitly designed tasks that elicited preservice teachers' beliefs and supported their understanding of mathematical content knowledge, our preservice teachers developed specialized curricular and pedagogical mathematics knowledge.

References

Arcytech. (2003). *Money*. Retrieved October 25, 2008 from http://arcytech.org/java/money/money.html.

Ball, D. L. (2003, February). *What mathematical knowledge is needed for teaching mathematics?* Paper presented at the Secretary's Summit on Mathematics, Washington, DC.

Ball, D. L., & Cohen, D. K. (1996). Reform by the book: What is - or might be - the role of curriculum materials in teacher learning and instructional reform? *Educational Researcher*, *25*(9), 6-8, 14.

Ball, D. L., & Cohen, D. K. (1999). Developing practice, developing practitioners: Toward a practice-based theory of professional education. In

G. Sykes & L. Darling-Hammond (Eds.), *Teaching as the learning profession: Handbook of policy and practice* (pp. 3-32). San Francisco: Jossey Bass.

Battey, D., Kafai, Y., & Franke, M. (2005). Evaluation of mathematical inquiry in commercial rational number software. In C. Vrasidas & G. Glass (Eds.), *Preparing teachers to teach with technology* (pp. 241-256). Greenwich, CT: Information Age Publishing.

Cameron, A., Dolk, M., Fosnot, C. T., Hersch, S. B., & Werner, S. (2006). *Young mathematicians at work: Minilessons for operations with fractions, decimals, and percents, grades 5-8*. Portsmouth, NH: Heinemann.

Darling-Hammond, L. (1998). Teacher learning that supports student learning. *Educational Leadership, 55*, 6-11.

Franke, M., & Chan, G. (2006) *High leverage practices*. Retrieved June 1, 2007 from: http://gallery.carnegiefoundation.org/insideteaching/quest/megan_loef_franke_and_angela_grace_chan_high.html

Freudenthal Institute. (2008). *Number line*. Retrieved October 25, 2008 from http://www.fi.uu.nl/toepassingen/03106/task1.html.

Hjalmarson, M. A. (2005). Purposes for mathematics curriculum: Preservice teachers' perspectives. In G. M. Lloyd, M. R. Wilson, J. L. M. Wilkins, & S. L. Behm (Eds.), *Proceedings of the 27th annual meeting of the North American Chapter of the International Group for the Psychology of Mathematics Education* [CD-ROM]. Eugene, OR: All Academic.

Lampert, M., & Ball, D. L. (1998). *Mathematics, teaching, and multimedia: Investigation of real practice*. New York: Teachers College Press.

Lloyd, G. M. (2002). Reform-oriented curriculum implementation as a context for teacher development: An illustration from one mathematics teacher's experience. *Professional Educator, 24*(2), 51-61.

Lloyd, G. M., & Behm, S. L. (2005). Preservice elementary teachers' analysis of mathematics instructional materials. *Action in Teacher Education, 26*(4), 48-62.

National Council for Teachers of Mathematics. (2008). *100's board and calculator*. Retrieved October 25, 2008 from http://standards.nctm.org/document/eexamples/chap4/4.5/calc_full/standalone2.htm.

National Library of Virtual Manipulatives. (2008a). *Base blocks addition*. Retrieved October 25, 2008 from http://nlvm.usu.edu/en/NAV/frames_asid_154_g_3_t_1.html?from=category_g_3_t_1.html.

National Library of Virtual Manipulatives. (2008b). *Chip abacus*. Retrieved October 25, 2008 from http://nlvm.usu.edu/en/nav/frames_asid_209_g_2_t_1.html?open=activities&from=category_g_2_t_1.html.

National Research Council. (2001). *Adding it up: Helping children learn mathematics*. Washington, DC: National Academy Press.

Remillard, J. T. (2000). Can curriculum materials support teachers' learning? Two fourth-grade teachers' use of a new mathematics text. *Elementary School Journal, 100*(4), 331-350.

Remillard, J. T. (1999). Curriculum materials in mathematics education reform: A framework for examining teachers' curriculum development. *Curriculum Inquiry, 29*(3), 315-342.

Shulman, L. S. (1987). Knowledge and teaching: Foundations of the new reform. *Harvard Educational Review, 57*(1), 1-22.

Shulman, L. S. (1986). Those who understand: Knowledge growth in teaching. *Educational Researcher, 15*(2), 4-14.

Wilson, S. M., & Berne, J. (1999). Teacher learning and acquisition of professional knowledge: An examination of research on contemporary professional development. *Review of Research in Education, 24*, 173-209.

Margret Hjalmarson is an Assistant Professor of Mathematics Education at George Mason University. She earned her Ph.D. in Mathematics Education and M.S. in Mathematics from Purdue University in 2004 and 2000, respectively. Her research interests include mathematics curriculum, teacher professional development, and assessment in engineering and mathematics.

Jennifer Suh is an Assistant Professor of Mathematics Education at George Mason University. She received her Ph.D. in Mathematics Education Leadership from George Mason in 2005 and her M.A.T. from the University of Virginia in 1994. Her research interests include lesson study, representational fluency, and diverse student populations.

Steele, M.
AMTE Monograph 5
Inquiry into Mathematics Teacher Education
©2008, pp. 109-118

10

Shifting Roles, Shifting Perspectives: Experiencing and Investigating Pedagogy in Teacher Education

Michael D. Steele
Michigan State University

A long-held tenet in teacher education is that teacher educators should model the sorts of pedagogy that we would like our teachers to enact with their own students. What beginning teachers might appropriate from this modeling is an open question. In this chapter, I describe a teacher education experience in which I invited preservice teachers to critically analyze my pedagogy. From the classroom discourse, I identify two key moves in the discussion of modeled pedagogy: abstracting generalized principles about teaching and learning from the instructor's examples, and connecting teachers' experiences as learners to their work in the classroom.

There is strong consensus that teacher educators should model the types of student-centered instructional practices that they want teachers to enact in their own classrooms (e.g., Ball & Cohen, 1999; Smith, 2001; Wilson & Ball, 1996). The *Professional Teaching Standards* (National Council of Teachers of Mathematics [NCTM], 1991) state that modeling pedagogy can help teachers "develop ideas about what it means to teach mathematics, beliefs about successful and unsuccessful classroom practices, and strategies and techniques for teaching particular topics" (p. 127). Yet it is not always clear what, if anything, teachers appropriate from their exposure to this modeling and how these experiences might impact teachers' classroom practice. Teachers might simply mimic the pedagogical moves that a teacher educator modeled without considering the rationale or the conditions for use. Often times, surface-level appropriations lead to poor results, and teachers abandon student-centered teaching for a more familiar didactic approach (e.g., Grant, Hiebert, & Wearne, 1998). So, how can teacher educators help teachers appropriate models of effective pedagogy in ways that are useful in their own classroom? In this chapter I describe one attempt to aid teachers in appropriating models of effective pedagogy by engaging them in a teacher education experience, then stepping back to examine and analyze the pedagogy of the teacher educator.

Making the Invisible Visible

Observing teaching is a common practice in education. Preservice teachers approach their training already indoctrinated in an apprenticeship of observation (Lortie, 1975) – they know what teaching looks like based on their own experiences as students, but are not privy to the thinking behind particular pedagogical decisions. University-based experiences are designed to provide the rationale, but can be too distant from the actual work of teaching, particularly when those experiences embody a theory-into-practice approach that has been pervasive in teacher training programs (e.g., Ball & Cohen, 1999; Smith, 2001). The theory-into-practice approach presents theories or generalities about thinking and learning, such as constructivism or discourse-based teaching, frequently leaving preservice teachers to determine how to apply those ideas in the day-to-day work of classroom teaching, a task which has proven difficult across a wide variety of subjects and contexts (Clift & Brady, 2005). In contrast, practicing teachers have a wealth of teaching experience, but the closed-door culture that overwhelmingly prevails in American school systems limits the opportunities that teachers have to observe and discuss practice (Kardos & Johnson, 2007). For preservice and practicing teachers alike, it is unclear what tools teachers might possess in order to make meaningful connections between observed practice, theories of teaching and learning, and their own teaching.

One way for a teacher educator to attempt to bridge this gap is by modeling effective pedagogical practices with a group of teacher learners, and then inviting them to reflect on those practices as learners and teachers. Having teachers engage in and critically reflect on learning experiences has a number of potential benefits beyond simply observing effective teaching. Such experiences provide teachers an opportunity to understand the rationale and principles behind the particular teacher moves, and to connect those particulars to generalities about teaching. Additionally, teachers gain unique insight into the impact of pedagogical decisions by reflecting on teaching from both the learner's and the teacher's perspective.

The following excerpts emerged from the teaching of a Master's-level advanced mathematics methods course focused on middle grades geometry and measurement. Course participants were 25 preservice and practicing teachers representing a range of grade levels K-12. The goals of the course were to develop both content knowledge related to geometry and measurement and pedagogical knowledge that supports inquiry-oriented teaching. The course was practice-based (Ball & Cohen, 1999), in that the authentic tasks of teaching practice (e.g., mathematical tasks, narrative and video cases of teaching, student work) were central to all work in the course and were positioned as artifacts of teaching for investigation and inquiry.

All 12 class sessions over the six-week course were videotaped. I taught the class and maintained a reflective diary in which I recorded entries before and after each class session. For this investigation, I examined the videotapes and my reflective diary to identify course activities that included discussions of my pedagogy as a teacher educator. I then closely examined the nature of the discourse for each activity to identify themes. In the excerpts that follow, teachers are seen making two important moves in their discussions of my pedagogy: abstracting generalized principles about teaching and learning from the instructor's examples; and connecting their own experiences as learners in the course to their work as classroom teachers.

Theme 1: The Need for Generalized Principles

A key goal in any practice-based experience in which teaching is examined is to move from the particulars of the teaching artifacts to generalities about teaching (Stein, Smith, Henningsen, & Silver, 2000). Although teachers may recognize similarities or differences between classroom artifacts and their own teaching, this recognition does not automatically lead to changes in teacher practice. For meaningful change to occur, "teachers must learn to recognize events as instances of something larger and more generalizable" (Stein, Smith, Henningsen, & Silver, 2000, p. 34). One way to move teachers from particulars to generalities about teaching is to engage them in a specific learning experience and discuss the pedagogical moves made by the teacher educator that supported their own learning and how these moves can generalize beyond the specific situation. In this way, the learning experience also serves as a case of teaching and learning for teachers to examine and analyze.

In Class 2, teachers solved a mathematical task related to linear and square units and discussed a narrative case in which middle school students were engaged in a discussion of the same task. Following these activities, I asked teachers to identify moves that I had made to promote discussion of the task and the case. My goal for this discussion was for the teachers to identify my use of *revoicing* (O'Connor & Michaels, 1996), a technique for organizing discourse and flagging important understandings, and to begin to consider how and when revoicing could be used more broadly in their own teaching. The excerpt that follows shows teachers identifying the idea of revoicing in my pedagogy and moving towards generalities about how and when revoicing might be used with their own students. (Note that I am identified as MDS in these episodes.)

1 MDS: So how did I promote discussion during our work on
2 the task?
3 Uma: You're very good at echoing back what we say. To
4 make sure that you are understanding what we're
5 actually saying, so that we can say – 'well, no, that's
6 not what I meant, I meant THIS.' It gives the person
7 time to clarify, sometimes in their own head, what
8 came out of their mouth. You know what I mean?
9 'Ok so this is what I heard you say,' and that way it
10 gives somebody a chance to think 'yeah, that's what I
11 said' or 'no, it didn't sound that way in my head, I
12 have time to change it.' So echoing back what we
13 said.
14 MDS: So I hear a few things in what you're saying. [class
15 laughs] I hear that I'm echoing back what you said,

16 and it gives the person who offered the idea a chance
17 to agree or disagree. And I think there are ways you
18 can do that, and maybe this is something to think
19 about, there are ways you can do [it] that make that
20 opening more or less apparent. In terms of the student
21 saying 'well no, that's not what I thought.' Then it
22 also gives that person time to reflect on, 'is that really
23 what I thought?'. Does it [echoing back] serve any
24 other purposes? Maura?
25 Maura: Well, it can help you unpack what that particular
26 person said. If you repeat it, it helps you to
27 understand what they're saying…
28 MDS: Ivy, did you have something to add?
29 Ivy: I think in this class, not only for you to understand
30 what she's saying, but it gives us the opportunity to
31 hear it again, and maybe another way.
32 Noelle: I think there's also a difference between asking it
33 back as a question and saying it. When you ask it
34 back as a question, you're talking to the student and
35 the student is confirming what they said. But if you
36 say what the student said and rephrase it say, 'well
37 this is what Johnny said this is how you do it,' you're
38 kind of saying 'well he said it, but I'm gonna say it
39 better'…
40 (overlapping speech in which teachers debate Noelle's
41 assertion)
42 MDS: So by asking it back as a question, it keeps the
43 ownership where it originally resided. So one name
44 for this, that you might hear or many of you may
45 have already heard, is revoicing. And we call it
46 revoicing to contrast it with repeating, because what I
47 hear from what all of you are saying, is that it's a lot
48 more than repeating. It's crediting with ownership,
49 it's clarifying the thinking. And one thing that I'll
50 add that hasn't come up, is that it allows me, if
51 there's a particular part of the idea that I want to
52 highlight, or that I want to clarify or that I want to
53 add to, I can change the language and highlight that
54 point, or add something onto the end, in a way that
55 doesn't necessarily sound like me telling.

Class 2, Whole-Class Discussion

In opening up my pedagogy for examination, I had hoped that teachers would identify the particular moves I had made, and specifically the idea of revoicing. Uma (lines 3-13) talked about exactly this issue, mentioning that I "echoed back" and provided some additional detail about what that means. At this point, the particulars of my use of revoicing in the class had been made available in the public discourse for all teachers to consider. However, the fact that the move has been identified does not guarantee that it will now be useful to teachers in their own teaching. Uma's comment identified one function of revoicing – allowing the speaker to agree or disagree with the revoiced statement. I then pressed teachers to consider other dimensions of revoicing by asking if revoicing serves any other purpose (line 23-24). Through continued conversation, additional functions for revoicing were identified (e.g., lines 25-27, 29-31), conditions for use were debated (Noelle's comment in lines 32-33), and eventually a connection was made to how revoicing might be used in the teachers' own practice with students (lines 33-39). The teachers had begun to move past the particulars of how I used revoicing in their learning experience towards the more general characteristics of the move and how it might apply in their own classrooms. In this way, teachers have begun the shift from the particulars of my move to generalities about teaching.

Simply understanding the generalities of a pedagogical move does not guarantee that the move will be used in practice. Teachers also need to consider under what specific conditions one might use a pedagogical move with a group of students. In the same class, teachers had identified the particular use of wait time as an "effective" pedagogical move, and had made the general claim that wait time supported learning. The idea of wait time was one that I had hoped would arise; however, the notion that wait time was good did not necessarily provide teachers with insights into *how* wait time might be beneficial and *when* using wait time might support student learning. I pursued this issue by encouraging the teachers to discuss the conditions under which wait time might be most effective.

1	MDS:	I have a question related to this for everybody. Nancy
2		mentions wait time, and I think that's a phrase and a
3		concept that we're all familiar with. So I wonder a lot
4		about wait time… I wonder, is that always effective,
5		and under what conditions should I use wait time, or
6		not? (long pause)
7	Nick:	Are you using it now?
8	MDS:	Take 30 seconds and turn and talk to one of the
9		people next to you about this…
10		(*30 seconds pass*)
11		I usually honor you finishing your sentences, but
12		time's up. So wait time. Always good, always bad, it
13		depends, what do you think? Emily?
14	Emily:	I think it depends on the situation. If you are just
15		going over something and asking 'ok, well do you

16 have any other ideas on this?' -- they're just going to
17 sit there and wait for you to stop and go on to the
18 next thing if they don't want to talk anymore. But if
19 you've just presented them new information, or
20 something new, then it's important that you give
21 them enough time to go through the processes in their
22 mind for them to think of maybe something else
23 they'd like to say about it.
24 (several other teachers share thoughts)
25 MDS: What if I ask you, 'so did you like the case [we just
26 read], yes or no?'. And then I waited for 10 seconds.
27 So think about that question.
28 (10 seconds pass) What did you think about? Nancy?
29 Nancy: It's not a question that really takes that much thought.
30 I guess it could take a lot of thought, but we're
31 basically thinking, 'yes I liked it' or 'no, I didn't.'
32 Kelsey: Thinking about the Mathematical Tasks Framework,
33 like the part about implementation, when you
34 purposely build wait time into your lesson plan, that's
35 a cue to the kids that this is a high-level task, that this
36 isn't just memorization or review, this is something
37 important that you need to focus your energy on. Do
38 you know what I'm saying? Like it's a cue that you
39 really need to get your mind into this one and this is
40 important and that we're really going to pull a lot of
41 math out of this particular task.

Class 2, Whole-class discussion

In this case, teachers had identified the particular use of wait time and generalized the idea as a move that supported student learning. If the conversation had ended there, the conditions for using wait time in the classroom may not have been clear. Emily's comment in lines 14-23 began to address conditions for use, but her response was relatively vague and abstract. In response, I engaged teachers in a first-person experience with wait time (lines 25-28), which led to teachers identifying a specific set of conditions when wait time is useful in the classroom, evidenced by Nancy's (lines 29-31) and Kelsey's (lines 32-41) statements. Kelsey's contribution is particularly interesting, as she moved from the question that was modeled, *Did you like the case?*, to thinking about wait time with respect to mathematical tasks in her own planning and practice. In this discussion, teachers moved from the particulars of my practice to generalities about teaching, and further to identifying conditions for these generalities to be useful in their classroom practice.

Theme 2: Insight into Pedagogy as a Learner and Teacher

To plan lessons in which teachers consider deeply how students will make sense of the mathematics and how the pedagogy might support their learning, teachers must be able to consider the student perspective. Analyzing student work and examining narrative or video cases of teaching are both popular ways to move teachers towards thinking about the student perspective. Narrative and video cases, in particular, provide some clues as to how a teacher's pedagogical move impacts student thinking and learning. Analyzing the pedagogy of a learning experience in which they have been engaged is a particular type of case, and one that affords teachers first-person insights into the student perspective as well as offering an opportunity to explore the teacher perspective. Turning a reflective lens on the pedagogy that they have just experienced as learners enables teachers to make more causal links between the instructor's moves and their own thinking. Discussions of a teacher educator's pedagogy also afford teachers an opportunity to hear how learners with diverse backgrounds and experiences responded to the same instructional moves. The next excerpt followed a two-class activity sequence during which I asked teachers to consider the nature of proof, to examine and construct proofs of their own, and to examine the process by which they created and evaluated proofs.

1 MDS: So I'd like you to think back on our experiences
2 over the last 2 classes about proof, and I'd like you
3 to think, what did I do to support your learning
4 related to proof?
5 [*teachers take time to collect their thoughts*]
6 Nancy: Um, I just said that you made me think about what
7 proof is and it might be... so everyday I don't think
8 to myself, 'What is proof?' [but now] I've been
9 thinking, 'What is proof?'! You've just helped me
10 open my mind up and think about it more deeply.
11 Ed: You gave us all these proofs to look at, kind of put
12 us in a student's position where we're looking at
13 something on the board, which we may or may not
14 make sense of, and it kind of gave us the impression
15 that sometimes we may put what we call a proof up
16 on the board that we may understand and we say
17 'this this this this this, there's the proof we're done,'
18 and the kids in the class are going like this. [*shrugs*
19 *shoulders and makes facial expression indicating*
20 *lack of understanding.*] 'What in the heck is this guy
21 talking about?' So maybe a proof to us isn't exactly
22 a proof, in our students' mind. It may prove to them
23 that they may not understand a lick of what we're
24 doing.

25 Melinda: You asked the question every day, or every few
26 classes, but you gave us new information to go
27 along with that question. So every time you'd add
28 something new to it [and ask], 'what's proof now
29 that we've talked about this?'.
30 MDS: So in asking it over and over it wasn't just
31 repeating?-
32 Melinda: No. You would give us new information to add to
33 our understanding.
34 Kelly: I think just the selection of tasks really tried to
35 pinpoint what the criteria was for us to think about
36 at the time.
37 Uma: It seemed like when we were working in our groups
38 and you would walk around, you always had a good
39 question. You'd listen and then you would ask
40 something to make us either clarify what our
41 thoughts were or rethink what we were doing at the
42 time. So you kept us moving in that same direction.
43 Daulton: Also you just refused to tell us what you thought a
44 proof was.
45 MDS: And still do.

Class 7, Whole-class discussion

Two insights in particular stand out from this excerpt as evidence of teachers considering both the learner and the teacher perspective. In lines 11-24, Ed spoke about how he felt as a learner when asked to consider eight proofs of the Pythagorean Theorem and decide which were or were not proofs. He drew a direct comparison to how students might feel when a teacher presents an idea that seems intuitive to him or her, such as sharing a completed proof. His comments suggest that teachers need to consider how students might make sense of a mathematical idea, and what the teacher might do to support meaningful understanding of a complex idea such as proof.

Several of the other teachers (Nancy in lines 6-9, Melinda in lines 25-29, Kelly in lines 34-36, Uma in lines 37-42, and Daulton in lines 43-44) articulated insights that link the learner and the teacher perspective. All five teachers talked in some way about how the sequence of activities impacted their learning, but they also made links to particular pedagogical moves. Nancy identified asking a question, "What is proof?" that she had not previously considered; Melinda noted the same question, discussing how it was asked after a new experience that might afford a different perspective on proof; and Daulton discussed the fact that I refused to give a definitive answer, providing an open forum for continued inquiry. In this way, the teachers linked their experiences as learners to specific teacher moves and to the overarching design of the series of activities and my goals as the teacher educator.

In the Class 7 excerpt, teachers identified teaching moves I made that supported learning. But perhaps more importantly, they were able to articulate how the moves supported *their* learning. Treating the activities as a case of learning and teaching, they were able to reflect from a first-person perspective on their learning and how it was supported in ways that the analysis of a third-person case or of student work may not afford. Moving one step further, Ed's comment in line 11 spontaneously connects his experiences as a learner to his students' experiences as learners, suggesting that engaging in the analysis of the pedagogy he had just experienced may impact the pedagogy he uses in his classroom practice.

Conclusion

Practice-based teacher education attempts to address a critical need in the professional training of teachers – linking theoretical perspectives and classroom practice. Having teachers analyze materials, such as student work and narrative or video cases of teaching, helps them plan and enact lessons in ways that consider the learner's perspective and the teacher moves that support student learning. Narrative and video cases, in particular, show the relationship between pedagogy and learning, and often include insight into the teacher's thinking through narrative expositions or pre- and post-lesson interviews. Engaging teachers in a learning experience and inviting them to examine the teacher educator's pedagogy adds a dimension that the analysis of a third-person case may not afford. By taking part in a learning experience and analyzing the pedagogy of that experience, teachers are able to consider learning in the first person. Moreover, they interactively investigate the teacher educator's moves through questioning and discussion. This analysis activity allows teachers to move from the particulars of their own learning to generalities about teaching, and potentially invite reflection on their own practice as teachers. Finally, engaging a group of teachers in the examination of a teacher educator's pedagogy provides the teacher educator with important insights regarding his or her own teaching. By making facilitation transparent, both teachers and teacher educators are given the opportunity to reflect on their practice in ways that have the potential to foster growth in their teaching.

References

Ball, D. L., & Cohen, D. H. (1999). Developing practice, developing practitioners: Toward a practice-based theory of teacher professional education. In L. Darling-Hammond & G. Sykes (Eds.), *Teaching as the learning profession: Handbook of policy and practice* (pp. 3-32). San Francisco: Jossey-Bass.

Clift, R. T., & Brady, E. (2005). Research on methods courses and field experiences. In M. Cochran-Smith & K. M. Zeichner (Eds.), *Studying teacher education: The AERA panel of research and teacher education* (pp. 309-424). Washington, DC: American Educational Research Association.

Grant, T. J., Hiebert, J., & Wearne, D. (1998). Observing and teaching reform-minded lessons: What do teachers see? *Journal of Mathematics Teacher Education, 1*(2), 217-36.

Kardos, S. M., & Johnson, S. M. (2007). On their own and presumed expert: New teachers' experience with their colleagues. *Teachers College Record, 109(*9), 2083-2106.

Lortie, D. C. (1975). *Schoolteacher: A sociological study.* Chicago: University of Chicago Press.

National Council of Teachers of Mathematics. (1991). *Professional standards for teaching mathematics.* Reston, VA: Author.

O'Connor, M. C., & Michaels, S. (1996). Shifting participant frameworks: Orchestrating thinking practices in group discussions. In D. Hicks (Ed.), *Child discourse and social learning* (pp. 63-102). Cambridge: Cambridge University Press.

Smith, M. S. (2001). *Practice-based professional development for teachers of mathematics.* Reston, VA: National Council of Teachers of Mathematics.

Stein, M. K., Smith, M. S., Henningsen, M. A., & Silver, E. A. (2000). *Implementing standards-based mathematics instruction: A casebook for professional development.* New York: Teachers College Press.

Wilson, S. M., & Ball, D. L. (1996). Helping teachers meet the Standards: New challenges for teacher educators. *Elementary School Journal, 97*(2), 121-38.

Michael D. Steele is an Assistant Professor of Teacher Education at Michigan State University. His research interests include investigating mathematical knowledge for teaching and how preservice and practicing teachers acquire and make use of this knowledge in classroom practice. He currently supervises the secondary mathematics teacher preparation programs at MSU.

Cady, J., Hopkins, T., and Hodges, T.
AMTE Monograph 5
Inquiry into Mathematics Teacher Education
©2008, pp. 119-129

11

Lesson Study as Professional Development for Mathematics Teacher Educators

Jo A. Cady
University of Tennessee

Theresa M. Hopkins
Thomas E. Hodges
Tennessee Governor's Academy

Teacher educators often discuss Japanese lesson study with their students yet do not consider it for their own professional development. This chapter follows the experience of two teacher educators as they conducted a lesson study to improve their teaching of place value to preservice and inservice teachers. We describe our journey, from identifying the curriculum problem and the genesis of the idea to conduct a lesson study, through the actual lesson study process. We also discuss the impact of the experience on our professional growth as mathematics teacher educators.

Our struggles to deepen our preservice teachers' and inservice teachers' mathematical understanding in teaching place value brought us to examine ways in which we might improve our own teaching. As mathematics teacher educators, we need to encourage the development of robust knowledge not only in the practice of teaching but also in the basic foundations of elementary school mathematics. For that reason, in our classes and professional development, we attempt to move teachers away from teaching a 'narrow band of procedural skills' towards a more holistic instructional approach involving problem solving and the development of a conceptual understanding of mathematics. This type of teaching, however, requires that the teacher has a deep understanding of mathematics. As Ma (1999) states:

> [A] profound understanding of fundamental mathematics goes beyond being able to compute correctly and to give a rationale for computational algorithms. A teacher with profound understanding of fundamental mathematics is not only aware of the conceptual structure and basic attitudes of mathematics inherent in elementary mathematics, but is able to teach them to students. (p. xxiv)

We considered several possibilities for examining our own teaching practices. In the end, our prior experiences with Cognitively Guided Instruction (CGI) and its focus on students' thinking led us to investigate Lesson Study as a means to improve our instruction and our students' understanding. In the remainder of this chapter we provide a brief description of lesson study and the lesson study cycle, our experiences when we engaged in lesson study as mathematics teacher educators, and the lessons we learned from engaging in lesson study.

Lesson Study

"It [lesson study] is the linchpin of the improvement process" (Stigler & Hiebert, 1999, p. 111). Educators in the United States became interested in Japanese lesson study late in the 1990s, following the publication of results from the Third International Mathematics and Science Study (TIMSS; now called Trends in International Mathematics and Science Study). The TIMSS Video Study illuminated that differences exist not only in the mathematical achievement of American and Japanese students, but also the manner in which students are taught. Analyses of videotapes shot in 8th-grade mathematics classrooms indicated that in the United States "mathematics teaching is extremely limited, focused for the most part on a very narrow band of procedural skills" (Stigler & Hiebert, 1999, p. 10), while "students in Japanese classrooms spend as much time solving challenging problems and discussing mathematical concepts as they do practicing skills" (p. 11).

The differences in classroom instruction were attributed in part to the Japanese teachers' use of lesson study. Although lesson study began as a grassroots movement, it is now sanctioned by the Japanese ministry of education as one form of professional development for teachers (Fernandez, 2002). The school-based lesson study seen in Japan stands in sharp contrast to the traditional workshops led by outside experts that comprise the view of professional development held in the United States.

Our Lesson Study Experience

Our lesson study began during a conversation regarding lessons on place value. Two of us had just taught a lesson on place value in our university courses, using a base five number system. We thought that teachers' current understanding of the base ten system was based on memorized, procedural knowledge and that the use of base five would create a cognitive dissonance that would force them to examine more closely the ideas associated with place value. However, rather than work within the base-five system, many of our participants simply converted from base five to the more familiar base ten. Our discussion of this lesson led to a decision to use a Lesson Study model to create and examine a lesson focusing on improving preservice and inservice teachers' understanding and instruction of place value. (See Figure 1 for our Lesson Study Cycle.)

1. CHOSE TOPIC
Identified teachers' weaknesses in teaching place value. Researched important concepts and teaching methods for place value. How could we create cognitive dissonance? What are appropriate representations?

4. EVAULATED
What contributed to or hindered understanding of place value? What were the trouble spots? How could the lesson be improved? What questions did the researchers have that required more study?

Our Lesson Study Cycle

2. CREATED ORPDA TASKS
Anticipated trouble spots. How might teachers see the task? What questions would we ask? Analyzed the approach based on research in Step 1.

3. TAUGHT THE LESSON
Observers focused on teachers' mathematical understandings of place value, such as groupings and notations.

Figure 1. Our lesson study cycle. (adapted from Lewis, Perry, & Murata, 2006)

Using the lesson study cycle, we identified our first goal to be building connections between the key ideas of place value. For example, we wanted to connect the key idea of quantifying sets of objects by grouping by tens and treating these groups of ten as single units (Fuson, 1990; Steffe & Cobb, 1988) to the structure of the written notation that captures this information about grouping. In addition, we wanted teachers to connect and examine the different representations for quantities, such as written symbols and manipulatives, and to see the relationship between understanding and the ability to translate from one representation to another. An affective goal we identified was to foster teacher empathy for the struggles their students have with learning place value.

To meet these goals, we created a base-five system, Orpda, using symbols rather than numbers and following the same patterns as the base-ten system (see Figure 2). In our initial lesson, we introduced sets of objects (with quantities 1, 2, 3, and 4), the symbol we had created to represent each quantity, and the name for the symbol and quantity. After introducing all five symbols, we asked teachers to hypothesize how to use these Orpda symbols to represent the next set which contained five objects. Teachers suggested several possibilities. We accepted these and moved on to represent a set of six objects. Again, teachers suggested several possibilities. Teachers discussed how confusing the Orpda system would be unless we could find a unique symbolic representation for each quantity. Finally, one teacher suggested that a set of five objects should be represented by a two-digit symbol in which the first digit represents one group of five and the second digit represents no units. This suggestion provided an opportunity to discuss how the placement of the digits in a number determines its value in both the Orpda system and the base-ten system, and the use of zero as a placeholder. We challenged the participants to continue counting in Orpda and to find the corresponding symbol for each quantity. This initial lesson formed the basis for our research lesson. Readers interested in more detail about this lesson should read *What's the Value of @*#?* (Hopkins & Cady, 2007).

Quantity	Symbol	Word
🚲	*	star
🚲🚲	@	at
🚲🚲🚲	#	pound
🚲🚲🚲🚲	^	caret
🚲🚲🚲🚲🚲	*~	flub
🚲🚲🚲🚲🚲 🚲	**	doozle
🚲🚲🚲🚲🚲 🚲🚲	*@	sholt

Quantity	Symbol	Word
🚲🚲🚲🚲🚲 🚲🚲🚲	*#	pouflube
🚲🚲🚲🚲🚲 🚲🚲🚲🚲	*^	carflube
🚲🚲🚲🚲🚲 🚲🚲🚲🚲🚲	@~	atty
🚲🚲🚲🚲🚲 🚲🚲🚲🚲🚲 🚲🚲🚲🚲🚲	#~	poundy
🚲🚲🚲🚲🚲 🚲🚲🚲🚲🚲 🚲🚲🚲🚲🚲 🚲🚲🚲🚲🚲	^~	carety
🚲🚲🚲🚲🚲 🚲🚲🚲🚲🚲 🚲🚲🚲🚲🚲 🚲🚲🚲🚲🚲 🚲🚲🚲🚲🚲	*~~	skoobrat

Figure 2. Orpda symbols and names.

We taught the lesson a total of eight times. Each time we taught the lesson we met to debrief and discuss our elementary teacher/participants' understanding of the place value concepts we identified earlier. The remainder of this chapter focuses on step four, the debriefing process.

During each debriefing meeting, we reviewed our observation notes to examine the relationship between our instruction and our participants' understanding. The first question we wanted to answer was, "How can we modify the Orpda number system to enhance our participants' understanding of the structure of the written notation that captures information about grouping?" We also wanted to gather information that would help us make instructional decisions about, among other things, learning activities (which are most effective for developing grouping concepts?), sequence of instruction (what sequence of activities is most effective?), and supporting participants' growth in knowledge (how do we help our participants connect and examine the different representations for quantities, such as written symbols and manipulatives?).

Reflecting on the Mathematics

Our rationale for creating Orpda was to create cognitive dissonance for teachers and force them to examine place value relationships. We chose five symbols (creating a base-five system) on the keyboard to represent quantities, thinking that these symbols would be as abstract for our participants as the numerals (0-9) are for elementary students. The names we chose corresponded with the names of those characters. Although these were the only names we introduced in the initial lesson, the Orpda number system continued to evolve each time we presented the lesson. For example, during the first debriefing we realized we needed to create names for quantities over five since we were asking participants to count quantities to 25. This dilemma forced us to examine the patterns of the names of numbers used in the base-ten system. We created names that mimicked the base-ten system names. For example, the name of the quantities represented by one group of five and one or two units were words that did not follow an obvious naming pattern (in a similar manner to "eleven" and "twelve"). However, the quantities represented by one group of five and three or four units had a suffix just as base ten numbers do (i.e., –teen). For each iteration of the lesson, we became more efficient at presenting the Orpda number system and, consequently, had to invent names for larger quantities. We then followed the patterns of twenty, thirty, forty, etc. in base ten, in order to invent names for quantities that would represent two groups of five, three groups of five, and four groups of five. We stopped creating new names when we reached five groups of five (similar to ten tens, or one hundred, in the base-ten system). Thinking of names for quantities in the Orpda system forced us to examine more closely than we ever had before the patterns and relationships in the names of numerals in the base-ten system.

We were further challenged to represent these quantities using the symbols we had created. Originally, we used snap cubes to help us think about the symbols that would be appropriate and in keeping with the concepts inherent in the base-ten system. As we counted, we represented the quantity with snap cubes, grouping by fives. Thus, a rod of five cubes with no singles units required a two-digit symbol. The first digit represented one group of five and the second digit represented no units. Adding one more to this quantity and determining the symbol that would make sense was relatively easy until we got to the next group of five. Just as elementary students would do when thinking about going from nineteen (19) to twenty (20), we had to pause briefly and reflect about what would make sense. Two groups of five would also be represented by a two-digit symbol. The first digit was the symbol we had chosen to represent the quantity of two and the second digit was the numeral we had chosen to represent zero. Now the pattern was becoming obvious to us until we got to five groups of five. This would be similar to the base ten representation of ten tens or one hundred. Since we were now out of symbols, another place needed to be created to make this a three-digit numeral. Thus we had five groups with five in each group, or a square, similar to ten tens or one hundred in base ten. This quantity would be represented by the symbol we had chosen to represent the quantity one. The remaining two places would contain the symbol for zero, indicating no rods and no units were present. This thinking process helped us understand where our participants would likely struggle so we could create questions to help them connect the symbolic notation to the quantity and see the patterns in the notation.

One challenge that we did not foresee was the use of zero as a placeholder. We took for granted that teachers would be able to use the symbol we had chosen for zero as a placeholder to represent larger quantities. The first time we presented the lesson, a discussion ensued about the use of zero in the symbol to represent one group of five. Several teachers insisted that zero represented nothing and, therefore, could not be used to represent a different quantity. Even when other teachers pointed out the use of zero in 10, 20, 30, 100, 110, 207, they were hesitant to allow us to use zero as a placeholder. This discussion led us to investigate the history of zero and forced us to think about the pedagogical moves we were making in regards to zero.

Reflecting on Our Pedagogical Moves

The focus of the lesson was to develop participants' deep understanding of place value so they would be more successful when developing place value concepts with their elementary students. Two aspects of the lesson we believed were critical to develop understanding were creating cognitive dissonance and then progressing to the moment of participant understanding, which we called the "A-ha" moment. We knew we had created this cognitive dissonance when we read one participant's reflection:

> This class activity was a true eye-opener to how students struggle with the concepts of numbers and especially place value. Working in groups made the process a little less frustrating. Trying to grasp the concept of a foreign number system proved to be a more difficult process than I expected.

Once the teachers began focusing on the lesson and developing an understanding of the new system, we watched for "A-ha" moments. For most teachers, these moments were easy to recognize with verbal statements such as "I got that, too!" and "YES!." Non-verbal "A-has" included a range of actions, including self-satisfied nods and fist pumping. The goal throughout the lesson was to move each teacher from disequilibrium to the "A-ha" moments. A discussion about what generated the "A-ha" moment for the teachers and how this would translate to their elementary students' learning wrapped up the lesson.

With each iteration of the lesson, we also thought about the pedagogical moves that would force teachers/participants to think about the relationships and patterns in the system and connect these to the base-ten number system. The learning trajectory we intended was for teachers/participants to build connections among external representations (physical, symbolic, and written words) in an effort to promote construction of internal connections. We felt this trajectory would enhance their conceptual understanding of place value (grouping patterns, notation system, and multiple representations) and that external representations would be used as tools to record and demonstrate quantities and to communicate about quantities.

In the initial lesson, we had assumed that participants would look for and see the patterns in our notation system just as we had. Indeed, they also struggled with finding the correct notation for quantities when spanning "decades" or moving to the next period, just as we had when creating the number system. This perturbation is similar to the perturbation elementary students have when learning our base-ten system. Although snap cubes were available, most participants did not use them, even after we suggested they might help. Because the cubes helped us understand the notation system, we decided at our debriefing session to place more emphasis on the use of snap cubes to represent quantities and to relate the cubes to base ten blocks. For example, if participants were focusing on the patterns in the symbols as they were counting, we asked them to show us what the number would look like using the snap cubes and vice versa. We also inquired about the verbal or written name for the quantity in the set. In general, we wanted participants to become comfortable with different forms of representations and to build relationships between representations.

However, we found that most participants, even with prompting, were still not using the snap cubes but instead were relying on patterns they saw in the symbols. Our observations told us that those who used the cubes were more adept at finding the correct notation. In addition, these participants continually told us that the manipulatives were invaluable in developing their understanding. As one participant stated,

> I realized from this activity that manipulatives can be very useful when learning a number system. Every time I attempted to convert a number in our base-ten number system to its counterpart in the Orpda Number System, I heavily relied upon the pictures (of a cube, a flat, a rod, and a unit) I had drawn in order to determine how many of each I needed...In my opinion, children should not be without manipulatives when learning or exploring a number system.

As a result of this feedback, we insisted that all participants show a particular quantity with the cubes and "guided" them through a series of exercises relating quantities, symbols, and words. During the debriefing from this iteration, we decided that we were more directive and structured than we would like and we also decided that forcing teachers to use "our favorite representation" did not always develop the understanding we desired. Research on representations suggests that students be allowed to create and use their own representations (Lesh, Post, & Behr, 1987), which is consistent with our lesson study findings. However, they should also be presented with and asked to critique alternate representations. As a result, in the next iteration of the lesson we asked those who used manipulatives to show others how they used the manipulatives to solidify their understanding. We continued to encourage, but not require, the use of manipulatives in subsequent iterations of the lesson.

Each iteration of the lesson and debriefings led to other adaptations in our pedagogy, such as the order of the activities or the focus of the activity. For example, when we saw one teacher/participant write the symbols horizontally with five numerals in each row, we saw the patterns emerge, as they would in a base-ten hundreds chart. Thus, we created an Orpda chart and left many of the spaces blank. This gave us the opportunity to discuss the patterns in the Orpda chart and relate these patterns to the base-ten hundreds chart. We hoped that this activity would help teachers see the value in using the hundreds chart in their classroom to develop base-ten concepts and that they would have similar discussions about the base-ten patterns in the hundreds chart with their elementary students. As the lesson study progressed and we became more efficient at presenting the Orpda system, we added more activities that we felt would enhance participants' understanding. For example, we spent more time on oral counting and thus began counting with one object instead of with zero. In addition, we added activities focused on grouping from elementary mathematics methods books and adapted these activities to Orpda. We felt these grouping activities would encourage the use of manipulatives and help to connect the symbol to the quantity represented in the given set of objects.

Lessons Learned

"Lesson study is not just about improving a single lesson. It's about building pathways for ongoing improvement of instruction" (Lewis, Perry, & Hurd, 2004, p. 18). It is a process that enables teachers to examine their practice systematically in order to become more effective instructors (Fernandez & Chokshi, 2002). Thus, the intellectual processes that take place during lesson study are as important, if not more important, than the isolated products (i.e, lesson plans).

During our lesson study, we discovered that our own understanding of the patterns, groupings, and notations of our base-ten system became more explicit. In addition, the process enhanced our understanding of the teaching of place value to preservice and inservice teachers. The debriefing sessions forced us to reflect upon our practice, in particular the relationship between our instruction and our participants' understanding. Through the iterative process of testing and refining our ideas, we scrutinized the teaching and learning process and were forced to make explicit our rationale for each pedagogical decision. Thus, we were making sense of our practice through our discussions. After each iteration of the lesson, we researched effective ways to teach place value, other mathematical topics that had been introduced during post-lesson debriefings, or effective pedagogical decisions. We used the ideas from research, from our lesson observations, and from our participants' reflections to refine or expand the lesson.

As a result of our lesson study experience, our collegial interactions evolved. The most lasting impressions concerned the benefits of having another professional in our classroom. Thus, we enjoyed the benefits of the proverbial "two heads are better than one." As our lesson study evolved, we decided that the "observer" would question teachers about their thinking as the lesson progressed, thus becoming a participant/observer, and more aligned with our CGI background. In the Japanese model, observers in the classroom do not interact with the students, instead their focus is on observing the lesson. In addition, during the lesson we tried to find a moment where we could share the results of this questioning with each other and adapt the lesson as it progressed. We found this communication led to richer discussions and helped us relate the participants' experiences with the Orpda system to the development of base-ten understanding in their elementary school students.

Another benefit to our collegial interaction was the nature of the professional relationship between us. During the lesson study process, we built a level of trust that expanded our discussions from our teaching practice to our writing and other aspects of our work. We learned to take advantage of each other's strengths in analyzing and teaching the lesson and how to challenge each other in a non-threatening manner. As our trust in each other grew, it became easier to make suggestions about how the lesson should progress, about activities to use during the lesson, about questioning techniques, etc. without fear of being judged or having our ideas dismissed. We found we could let go of being right and critically reflect upon our decisions to create an effective lesson.

We also discovered that the lesson study process is time intensive. Individually, we needed time to read the literature about place value, representations, and effective instruction. Additional time was needed to share research, plan the initial lesson, teach or observe the lesson several times during the year and a half, discuss observations, and make adjustments to the lesson. For a successful lesson study, both of us committed large amounts of time.

Conclusions

We felt confident that our goal to create a lesson that provided teachers/participants with new knowledge about place value and students' struggles with place value was met when one teacher/participant stated:

> The Orpda number system has helped me realize the value of acquiring number sense before "blindly" following a procedure. Traditional teaching methods that emphasize rote memorization of facts assume that students will learn the concepts by practicing the

> algorithm over and over...I now understand even more that thinking and grappling with problems and concepts takes time. And some students (like me) need to have access to concrete representations longer than the teacher (like me) would necessarily think.

In addition, working together provided us with more insight and ideas than either would have had individually, creating a richer lesson and classroom experience. For other mathematics teacher educators considering lesson study as professional development, we offer the following for you to consider: select a critical topic; work with someone you trust and respect; and conduct reflection sessions shortly after each lesson while the observations are fresh. When choosing a lesson study topic, remember that the focus is a lesson, not an entire course.

Engaging in a lesson study as mathematics teacher educators increased our own pedagogical content knowledge of the teaching of place value to preservice and inservice teachers. The reflections and observations inherent in lesson study increased our conviction of the value of collegial interaction and the value of focusing on teacher/participants' understanding as a way to support their deep mathematical understanding. Lesson study provided an opportunity for further collaboration and professional growth in areas besides teaching. For example, it has extended to publications and discussions about research and theories.

Our participants also grew in their understanding of place value concepts as well as the need to focus and reflect upon their students' thinking throughout the lesson. We knew our affective goal had been met when one teacher/participant so aptly stated:

> "Walk a mile in my shoes" is a phrase we often say to others when we want them to experience something from our point of view. Putting myself in the role of a student learning the concept of place value was just that - walking in a student's shoes. I loved the experience. As a teacher I forget sometimes how hard it is for students when they first encounter math concepts that I take for granted.

References

Fernandez, C. (2002). Learning from Japanese approaches to professional development: The case of lesson study. *Journal of Teacher Education, 53*, 393-405.

Fernandez, C., & Chokshi, S. (2002). A practical guide to translating lesson study for a U.S. setting. *Phi Delta Kappan, 84*(2), 128-134.

Fuson, K. C. (1990). Issues in place-value and multidigit addition and subtraction learning and teaching. *Journal for Research in Mathematics Education, 21*(4), 273-280

Hopkins, T. M., & Cady, J. A. (2007). What's the value of @#^*? *Teaching Children Mathematics, 13*, 434-437.

Lesh, R., Post, T., & Behr, M. (1987). Representations and translations among representations in mathematics learning and problem solving. In C. Janvier (Ed.), *Problems of representations in the teaching and learning of mathematics* (pp. 33-40). Hillsdale, NJ: Lawrence Erlbaum.

Lewis, C. C., Perry, R., & Hurd, J. (2004). A deeper look at lesson study. *Educational Leadership, 61*(5), 18-22.

Lewis, C. C., Perry, R., & Murata, A. (2006). How should research contribute to instructional improvement? The case of lesson study. *Educational Researcher, 35*(3), 3-14.

Ma, L. (1999). *Knowing and teaching elementary mathematics: Teachers' understanding of fundamental mathematics in China and the United States.* Mahwah, NJ: Lawrence Erlbaum Associates.

Steffe, L. P., & Cobb, P. L. (1988). *Construction of arithmetical meanings and strategies.* New York: Springer-Verlag.

Stigler, J. W., & Hiebert, J. (1999). *The teaching gap: Best ideas from the world's teachers for improving education in the classroom.* New York: The Free Press.

JoAnn Cady is an elementary and middle school mathematics teacher educator. She works with preservice and inservice teachers regarding the implementation of practices that focus teachers' instructional decisions on their assessment of students' mathematical understanding. Her research interests focus on the development of pedagogical content knowledge and teacher beliefs.

Theresa M. Hopkins (PhD Teacher Education, University of Tennessee), teaches mathematics at the Tennessee Governor's Academy for Mathematics and Science. Her research interests include the use of reflection to improve teaching, place value, and gender issues in mathematics. She enjoys teaching mathematics and facilitating teacher workshops.

Thomas Hodges is a mathematics teacher at the Tennessee Governor's Academy for Mathematics and Science and a mathematics education instructor at The University of Tennessee in Knoxville, TN. He is interested in teacher identity and professional development.

Watanabe, T., Takahashi, A., and Yoshida, M.
AMTE Monograph 5
Inquiry into Mathematics Teacher Education
©2008, pp. 131-142

12

Kyozaikenkyu: A Critical Step for Conducting Effective Lesson Study and Beyond

Tad Watanabe
Kennesaw State University

Akihiko Takahashi
DePaul University

Makoto Yoshida
William Paterson University

As lesson study becomes more widely practiced in the United States, it is important that lesson study practitioners shift their focus from simply practicing lesson study to practicing lesson study effectively and meaningfully. In this chapter, we describe an important process in lesson study called kyozaikenkyu. We argue that a deep and critical kyozaikenkyu is an essential component of successful lesson study. We also discuss some implications for mathematics teacher educators in their work with both preservice and inservice teachers.

> Lesson study is not just about improving a single lesson. It's about building pathways for ongoing improvement of instruction. (Lewis, Perry, & Hurd, 2004, p. 18)

Since its introduction in the United States in the 1990s (Lewis & Tsuchida, 1998; Stigler & Hiebert, 1999; Yoshida, 1999), lesson study has attracted much attention among U.S. mathematics educators. Many mathematics teacher educators facilitate lesson study. However, as more U.S. teachers engage in lesson study, it is important that lesson study practitioners shift their focus from simply practicing lesson study to practicing lesson study effectively and meaningfully (Fernandez, 2002; Yoshida, Takahashi, & Watanabe, 2003). As Lewis' quote suggests, lesson study is about the improvement of mathematics instruction, not just the improvement of a single lesson. By planning, teaching, and discussing a publicly taught lesson, lesson study practitioners may learn many things. But for ongoing improvement of instruction, this learning must be purposeful and intentional; it must be about mathematics, mathematics learning, and mathematics teaching. In our experiences with teachers engaged in lesson study, we have found that they often do not attend well to an important step in the process called *kyozaikenkyu*. In this chapter, we describe this critical step of lesson study and argue for its importance. We also discuss some implications for mathematics teacher educators in their work with both preservice and inservice teachers.

Lesson Study and Improving Mathematics Instruction

Lewis, Perry, & Hurd (2004) pointed out the importance of looking beyond the visible features of lesson study. They identified seven "key pathways to improvement that underlie successful lesson study" (p. 19). We wholeheartedly agree that if lesson study's "visible features are implemented ritualistically, without a clear grasp of how they relate to instructional improvement" (p. 19), then we will not be able to take full advantage of this promising professional development practice. For example, a lesson study group may decide to use the problem shown in Figure 1 in their research lesson because it seems to offer a variety of solution methods. Problems like this one are often found in textbooks (e.g., Clements, Jones, & Moseley, 1999). However, why would we want students to find the area of this type of shape? Is it simply an application of area of rectangles? Why is it included in this lesson, which happens immediately before the lesson on area of right triangles and parallelograms? Unfortunately, the teachers' manual is silent on these questions. Without answering these questions, teachers might simply assign the problem; if their students can answer it correctly, perhaps using a variety of methods, teachers will happily proceed to the next lesson. However, what did the students learn? What did the teachers learn about teaching mathematics more effectively? The study group may have followed the visible features of lesson study closely, but they are not getting the most out of lesson study.

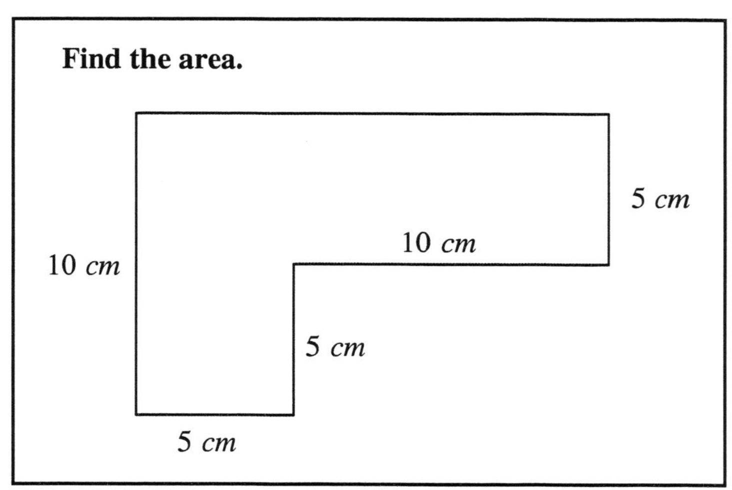

Figure 1. Problem adapted from Clements et al. (1999).

One way to avoid ritualistic or naive implementation of visible features is to gain a deep understanding of these visible features. In particular, lesson study practitioners need to understand what is involved in the planning of a research lesson. When Japanese teachers begin their planning, they initially engage in a practice called *kyozaikenkyu*, literally *study of instructional materials.* This practice is a central activity in teachers' everyday practice, but it plays a particularly important role in lesson study. In fact, one way lesson study contributes to the improvement of everyday instruction is through *kyozaikenkyu.*

What Are *Kyozai* (Instructional Materials)?

The word *kyozai* is written with two Chinese characters. The main meaning of the first character, *kyo*, is to teach or instruct, while the second character, *zai*, means materials (as in raw materials for manufacturing). Thus, the literal translation of the term is "instructional materials." When the phrase, "instructional materials," is used in the U.S. context, it tends to mean textbooks, or other prepared curriculum materials. However, the Japanese word *kyozai* means much more. According to *Jugyokenkyu Yougo Daijiten (Dictionary of Lesson Study Terms)* (Yokosuka, 1990),

> *Kyozai* is an actualization of educational content and includes learning goals. It is important that *kyozai* and subject matter content (specific knowledge and procedures to be learned through lessons) are distinguished. It is possible to explore the same subject matter with different *kyozai*, or we can investigate different subject matter with the same *kyozai*. (p.19, all translation in this paper was done by the first author)

The same dictionary states,

> To distinguish *kyozai* and subject matter content means to clearly articulate what specific subject matter content will be (can be) taught through a particular *kyozai*. If this relationship is not clearly articulated, the selection of *kyozai* or its interpretation becomes unclear. That, in turn, may make the purpose of the lesson unclear. (p. 19)

Thus, it appears that *kyozai* includes tasks or problems (with their contexts) and instructional tools. Implicit in this description is that to study instructional materials intensively means to articulate clearly the relationship between various *kyozai* and the content of a particular subject matter, or the goal understandings.

Speaking of "tools," there is a related Japanese term, *kyogu*. The second character in this term, *gu*, means tools. The same dictionary presents two contemporary interpretations of the term:

> (1) *kyogu* is concretized *kyozai* and includes such items as films, maps, models, scientific equipment, etc., and it can be considered as "non-linguistic *kyozai*." (2) *kyozai* and *kyogu* are the two faces of the same medium – *kyogu* is an outward face while *kyozai* is its inward face. This relationship is analogous to hardware and software of computers. (p.20)

Thus, another aspect of studying instructional materials intensively involves the study of instructional tools and their potential for helping students understand a particular subject matter's content.

Relationship Between Textbooks and *Kyozai*

The preceding discussion makes it clear that *kyozai* (instructional materials) includes much more than textbooks. However, research indicates that textbooks do play a central role in schooling, for both teachers and students (e.g., Ma, 1999, Schmidt, McKnight, & Raizen, 1996; Shimahara & Sakai, 1995). So, studying textbooks is an important part of the *kyozaikenkyu* process.

Textbooks are a collection of *kyozai*. Content standards, such as the *Principles and Standards for School Mathematics* [PSSM] (National Council of Teachers of Mathematics [NCTM], 2000) or state-level content standards or frameworks, simply list what students need to understand. They do not always suggest ordering or relative emphasis among content in particular grade levels (or particular courses like Algebra I). Textbooks are an interpretation of these standards; they present problems and tasks for studying various content using different instructional tools. Moreover, textbooks present these contents in a particular order, with suggested pacing. Teacher manuals accompanying textbooks further clarify what should be emphasized, how the current topic relates to what came before and will come later, rationale for *kyozai* and tool selection, etc. Clearly, because U.S. textbooks are published nationally in the absence of a national curriculum, the "fit" between textbooks and the "standards" is much less than what we might observe in other countries where there is a national curriculum. Moreover, because many U.S. textbooks are written in such a way that they can be used in different states with different content standards, teachers' manuals do not (cannot) always discuss the content relationships in detail. (For a comparison of U.S. and Japanese elementary mathematics teachers' manuals, see Watanabe (2001).) Therefore, the need for *kyozaikenkyu* is particularly important in the United States if textbooks are to be used more effectively.

Teachers must keep in mind that their responsibility is not to teach the textbooks, but rather to teach mathematics *with* textbooks. Textbooks are a collection of *kyozai*. Although the authors of a textbook series may have carefully thought about specific mathematics and selected appropriate materials and tools to teach that topic, teachers must understand why the particular *kyozai* was developed in order to take full advantage of *kyozai* in the textbooks. To maximize students' learning with the *kyozai* in any textbook, teachers must understand the true intent of the textbook authors. *Kyozaikenkyu* is the process to help teachers gain a deep understanding of *kyozai*.

What Is *Kyozaikenkyu* (Instructional Material Study)?

> *Jugyokenkyu Yougo Daijiten* (Yokosuka, 1990) defines *kyozaikenkyu* as the entire process of research activities related to *kyozai*, beginning with the selection/development, deepening the understanding of the true nature of a particular *kyozai*, planning a lesson with a particular *kyozai* that matches the current state of the students, culminating in the development of an instructional plan. (p. 73)

There are two types of *kyozaikenkyu* described in this dictionary. The first is the study of materials that are already developed as *kyozai*. In this process, the focus is whether or not the specific pre-developed material may be suitable in the teacher's classroom. The second type is actually done to develop *kyozai* so that the main emphasis appears to be an in-depth investigation of the particular subject matter.

Hayashi (1977 – as referenced in Yokosuka (1990)) suggested two stages of *kyozaikenkyu*. In the first stage, teachers need to investigate the *kyozai* deeply and repeatedly, until they clearly understand the nature of *kyozai*; teachers develop a clear vision of what they want to teach (or what they want students to understand) using this *kyozai*, and that vision provides the foundation of the lesson. In the second stage, teachers plan a lesson using a particular *kyozai*. In this process, teachers will think about how the task may be posed to the students, in what order, with what tools, etc.

During the second stage, teachers keep in mind that they cannot teach everything they learned in the first stage of *kyozaikenkyu*. There is a Japanese saying, "To teach one, you have to learn ten." In other words, once teachers "learn ten" during the first stage of *kyozaikenkyu*, they must critically evaluate all they learned and decide which ideas are essential for the lesson so that the lesson will have a clear focus. (Iwaasa, 1981, referenced in Yokosuka (1990)). Although teachers may not be directly using the vast majority of what they learned in the first stage of *kyozaikenkyu*, the deep understanding teachers gain in the first stage necessarily makes their understanding of the specific goal richer and more sharply focused.

Yoshida (1977 – as referenced in Yokosuka (1990)) suggested that *kyozaikenkyu* is the study of *kyozai* from a child's perspective, or an attempt to look at *kyozai* as children would and anticipate their responses and reactions to *kyozai*. Thus, although the first stage of *kyozaikenkyu* described by both Hayashi and Iwaasa seems to focus on understanding *kyozai* from a teacher's (adult) perspective, Yoshida expects teachers to ask themselves, "How does this look from my students' eyes?" based on what they currently understand.

Lesson Study and *Kyozaikenkyu*: Implications

A deep and critical *kyozaikenkyu* is an essential component of successful lesson study (Takahashi & Yoshida, 2004). Perhaps we can say lesson study offers an opportunity to conduct *kyozaikenkyu* intensively. So, what does this mean to our work as we support lesson study groups to conduct lesson study more effectively? One obvious implication is to help lesson study groups deepen the level of their *kyozaikenkyu*. As previously discussed, the process of *kyozaikenkyu* is intensive and complex. However, there are essential questions teachers need to ask in the process of *kyozaikenkyu* regarding the content of a specific subject matter. Those questions include, but are definitely not limited to:

- What does this idea really mean?
- How does this idea relate to other ideas?
- What is/are the reason(s) for teaching this idea at this particular point in the curriculum?
- What ideas do students already understand that can be used as a starting point for this new idea?
- Why is this particular problem useful in helping students develop this new idea?
- How can students solve this problem using what they already know, and how can their solution strategies be used to develop this new idea?
- What are common mistakes? Why do students make such mistakes? How should teachers respond to those mistakes?
- What new ideas are students expected to build using this idea in the future?
- What manipulatives and other materials should be provided to students? How do they influence students' learning?

It is essential that teachers understand the content for themselves as well as see the content from students' perspectives. To facilitate that inquiry, teachers should actually solve the potential problems and/or engage in the learning activities themselves. Teachers should not take these questions as simply a checklist. Rather, they must understand why these questions are important in planning a lesson.

An Illustration

We offer an illustrative example of *kyozaikenkyu* using the problem shown in Figure 1 (Adapted from Clements, Jones, Moseley, & Schulman, 1999). An important step in *kyozaikenkyu* is for teachers to ask themselves, "How would the students solve this problem?" Many students would likely calculate the area using the methods in Figure 2. Looking at these two solutions, teachers might consider one of the reasons to use this problem. In addition to determining the area of a composite shape made up of rectangles, the problem helps students understand a concise and efficient way to represent computation processes. But, how would this problem be a bridge to the next lesson on area of triangles and parallelograms? Are there other ways of finding the area of this figure that would lead into the next lesson? Moreover, what is it that we want students to understand? The more explicitly and clearly teachers can articulate the true purpose of the problem, the more likely that they can go beyond simply teaching how to solve the particular problem.

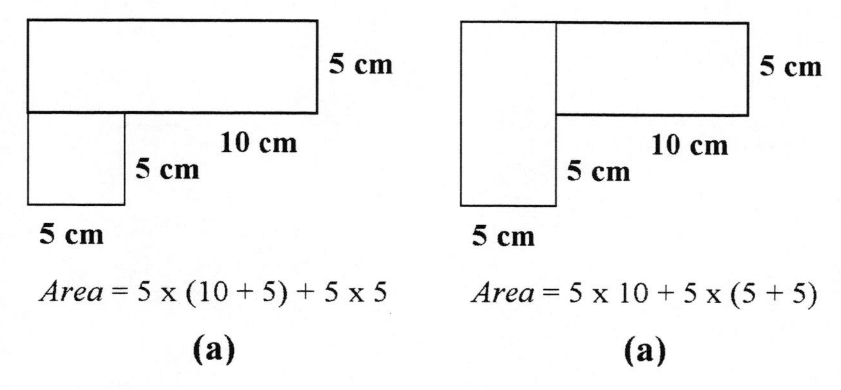

Area = 5 x (10 + 5) + 5 x 5

(a)

Area = 5 x 10 + 5 x (5 + 5)

(a)

Figure 2. Two possible ways to find the area.

Figure 3 shows four additional methods that might be used to calculate the area of the given shape. There are also other ways we might calculate the area. These approaches lead to another important question the study group might ask: "What might students learn from these solution methods?" After all, methods (a), (b) and (c) in Figure 3 will not work in general. So, some might question the value of these strategies. However, if the authors intentionally chose dimensions that allow these strategies to work, there may be some benefit for students to use these strategies. As teachers compare and contrast these methods, they might notice that one possible idea students could learn is that a given shape can be changed into a familiar shape for which there is an already learned method to calculate its area. Making this general principle explicit might be worthwhile. Furthermore, the methods shown in Figure 2 *partitioned* the given shape into parts which are familiar shapes while the methods shown in Figure 3 created familiar shapes by *transforming* the given shape. In addition, there are two particular transformations of the given shape that might be useful later: "cut and re-arrange" (Figures 3a and 3c) and "copy and re-arrange" (Figure 3b). For example, the cutting and re-arranging method might be useful as students try to determine the area of triangles and parallelograms as shown in Figure 4. In contrast, the copying and re-arranging method might be useful as students try to calculate the area of triangles or trapezoids (Figure 5).

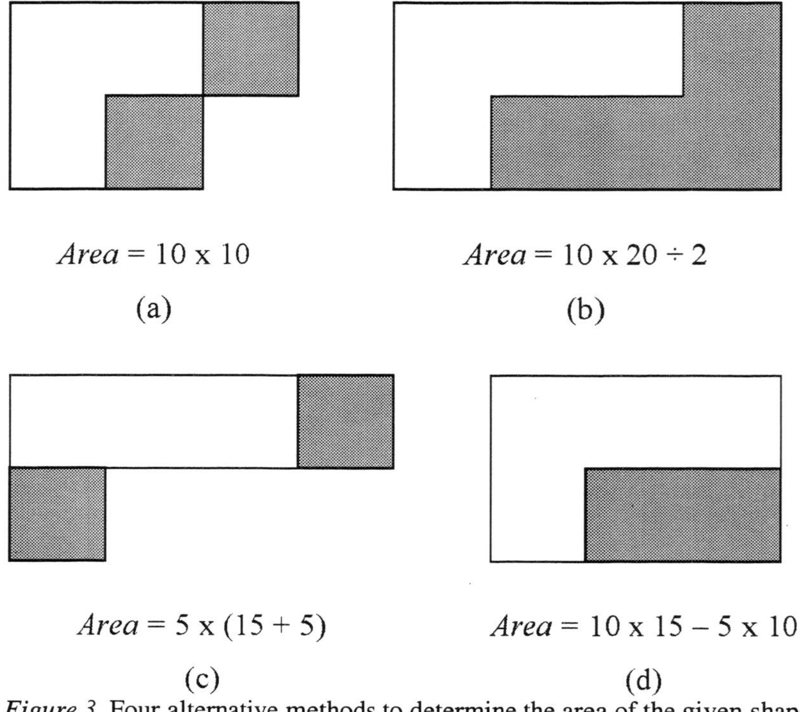

Figure 3. Four alternative methods to determine the area of the given shape.

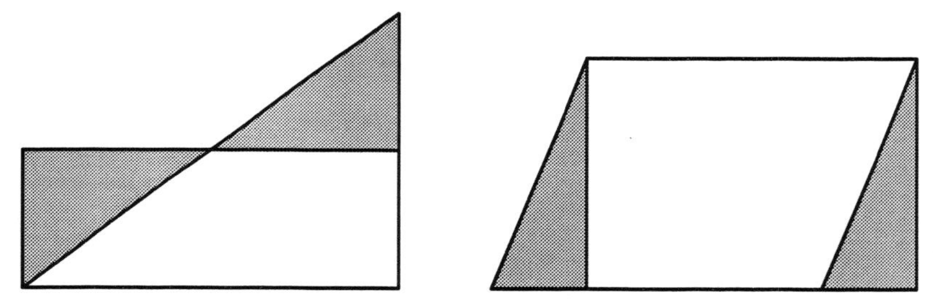

Figure 4. Determining the area of a right triangle and a parallelogram using the cutting and re-arranging method.

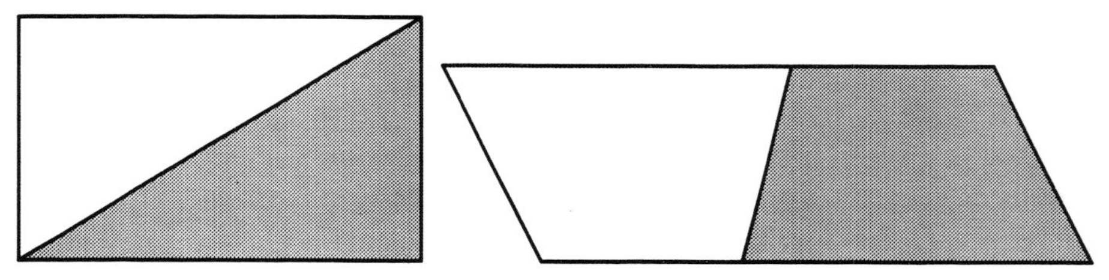

Figure 5. Using the copying and re-arranging method to calculate the area of a triangle and a trapezoid.

Thus, as a result of *kyozaikenkyu*, teachers might conclude that the "big idea" they want students to develop is that by transforming the given figure to something familiar students can determine its area. In fact, the goal of this problem is not just finding the area of the shape, but to understand how to use previously learned area formulas to find the area of unfamiliar shapes. The *PSSM* (NCTM, 2000) states that all students should "develop, understand, and use formulas to find the area" (p. 170) of various shapes. Therefore, students' understanding of the area-preserving transformation (cut and re-arrange) and the area-doubling transformation (copy and re-arrange) is critical if they are to *use* formulas to *develop* other area formulas.

If teachers want students to understand those methods, how should they pose the problem? This is another important question for the study group. How likely is it for any student to use the methods shown in Figure 3? The chances are slim if teachers simply ask students to determine the area of the shape. Thus, teachers may need to consider a different way to pose the problem to encourage students to focus on methods of determining the area of unfamiliar shapes. Based on these considerations, teachers might decide to pose the problem in the following way:

> Let's find several different ways we can use the formulas we have already learned to find the area of the following shape.

Moreover, questions like, "Is there a way to use the formula of rectangles (or squares)?" or "Can you divide the shape into two parts so that you can use what you already know?" may be useful to help those students who struggle with this problem.

Teachers should also consider how the problem is presented to students. Would it matter if the problem were given on a geoboard or on a piece of paper so that students could actually cut it out? Should the figure be drawn on a grid, or should the students cut out the figure and place it on separate grid paper? Which measurements should teachers provide? The problem includes 5 measurements – why not the measurements for all 6 lengths? Should teachers provide the measurements for only 4 lengths? As teachers consider these questions, they need to keep in mind the goal of this lesson – to help students learn how to use previously learned formulas to find the area of unfamiliar shapes.

Once students develop various methods to determine the area of this shape, teachers should be prepared to orchestrate the class discussion. In what order should these methods be shared? What might be an effective way to encourage students to compare and contrast the various methods? What might be the strengths of each method? Weaknesses? Teachers should think about these questions while keeping in mind the goal of the lesson – to find the area of an unfamiliar shape by changing it into a familiar shape and to provide experience with two specific methods (area-preserving and area-doubling transformations) for changing the given shape.

Concluding Remarks

Figure 6 summarizes the *kyozaikenkyu* process. Teachers engage in *kyozaikenkyu* everyday. Although teachers might not conduct an in-depth *kyozaikenkyu* for every single lesson, experiencing a deep *kyozaikenkyu* in lesson study will no doubt improve their everyday *kyozaikenkyu*. Moreover, *kyozaikenkyu* is the particular mechanism in lesson study that leads to the deepening of teachers' content knowledge. It is our hope that readers now have a deeper understanding of one specific visible feature of lesson study – planning a research lesson that will help them build "pathways for ongoing improvement."

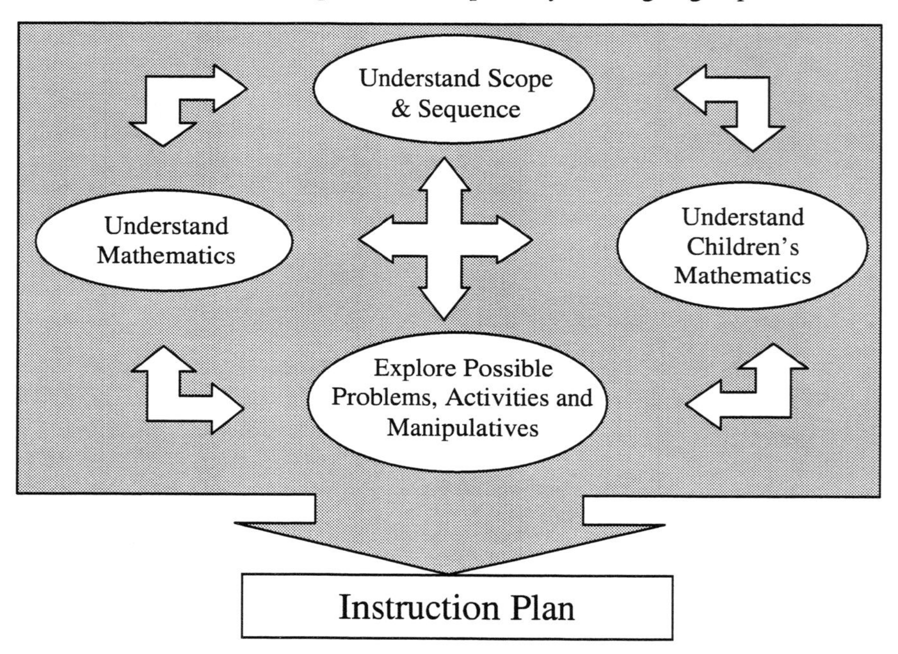

Figure 6. Kyozaikenkyu process.

Furthermore, we believe it is essential that lesson study groups share the results of their *kyozaikenkyu* with each other. Stigler and Hiebert (1999) pointed out the need for an easily accessible knowledge base of teaching for teachers; lesson study, perhaps, may serve as a means to establish this knowledge base. We believe that such a knowledge base should include more than a collection of exemplary lessons, but also information that helps teachers understand why they teach a particular idea and how it relates to other ideas. Moreover, such a knowledge base should not be simply a collection of "rules and formulas," but, rather, it should be a beginning for deeper *kyozaikenkyu* by other teachers.

Finally, the practice of *kyozaikenkyu* should not be occurring only in the context of lesson study. We believe it is essential that mathematics teacher educators engage in *kyozaikenkyu* of school mathematics themselves in order to gain deeper understandings of school mathematics. Furthermore, *kyozaikenkyu* is something that teachers must continually practice themselves, not just while engaged in lesson study. When teachers do not have a deep understanding of curriculum materials, even a high-quality curriculum could be implemented in a procedural way. Therefore, the courses mathematics teacher educators teach, whether content or pedagogy focused, should offer opportunities for preservice and inservice teachers to engage in the practice of *kyozaikenkyu*.

References

Clements, D., Jones, K., & Moseley, L. (1999). *Math in my world: Teacher's edition, Grade 5, Part 2*. New York: McGraw-Hill.

Fernandez, C. (2002). Reflections on implementing lesson study in the U.S. Presentation at the meeting of the 2002 Lesson Study Conference, Stamford, CT.

Lewis, C. C., Perry, R., & Hurd, J. (2004). A deeper look at lesson study. *Educational Leadership, 61*(5), 18-22.

Lewis, C., & Tsuchida, I. (1998). A lesson is like a swiftly flowing river: Research lessons and the improvement of Japanese education. *American Educator, 22*(4), 12-17, 50-52.

Ma, L. (1999). *Knowing and teaching elementary mathematics: Teachers' understanding of fundamental mathematics in China and the United States.* Mahwah, NJ: Erlbaum.

National Council of Teachers of Mathematics. (2000). *Principles and standards for school mathematics*. Reston, VA: Author.

Schmidt, W., McKnight, C., & Raizen, S. (1996). *A splintered vision: An investigation of U. S. science and mathematics education.* Boston: Kluwer.

Shimahara, N., & Sakai, A. (1995). *Learning to teach in two cultures: Japan and the United States.* New York: Garland.

Stigler, J., & Hiebert, J. (1999). *The teaching gap: Best ideas from the world's teachers for improving education in the classroom.* New York: The Free Press.

Takahashi, A., & Yoshida, M. (2004). Ideas for establishing lesson-study communities. *Teaching Children Mathematics, 10* (9), 436-443.

Watanabe, T. (2001). Content and organization of teachers' manuals: An analysis of Japanese elementary mathematics teachers' manuals. *School Science and Mathematics, 101*(4), 194-205.

Yokosuka, K. (1990). Jugyokenkyu yougo daijiten. (Dictionary of lesson study terms). Tokyo, Japan: Tokyo Shoseki.

Yoshida, M. (1999). Lesson study [Jugyokenkyu] in elementary school mathematics in Japan: A case study. Paper presented at the Annual Meeting of the American Educational Research Association, Montreal, Canada.

Yoshida, M., Takahashi, A., & Watanabe, T. (2003). Prerequisite for planning lessons effectively. Presentation at the meeting of the 2003 Lesson Study Conference, Stamford, CT.

Tad Watanabe is an Associate Professor of Mathematics Education at Kennesaw State University. His interests include teaching and learning of multiplicative concepts and mathematics education practices in Japan. He teaches mathematics content courses for prospective teachers at KSU, and he has worked with lesson study groups throughout the United States.

Akihiko Takahashi, Ph.D., is an Associate Professor of Mathematics Education and the Director of the Asia-Pacific Mathematics and Science Education Collaborative at DePaul University. He teaches mathematics content and methods courses for prospective teachers and provides workshops and seminars for practicing teachers using ideas from the U.S. and Asia.

Makoto Yoshida is the Director of the Center for Lesson Study at William Paterson University. He introduced lesson study from Japan and continues to educate teachers about it through conferences and workshops. He is particularly interested in working with mathematics teachers to improve their content knowledge and pedagogy through lesson study.

Neiss, M., Ronau, R., Driskell, S., Kosheleva, O., Pugalee, D., and Weinhold, M.
AMTE Monograph 5
Inquiry into Mathematics Teacher Education
©2008, pp. 143-156

13

Technological Pedagogical Content Knowledge (TPCK): Preparation of Mathematics Teachers for 21st Century Teaching and Learning

Margaret L. Niess
Oregon State University

Robert N. Ronau
University of Louisville

Shannon O. Driskell
University of Dayton

Olga Kosheleva
University of Texas at El Paso

David Pugalee
University of North Carolina at Charlotte

Marcia Weller Weinhold
Purdue University Calumet

Technological pedagogical content knowledge (TPCK) challenges teacher preparation programs to rethink the preparation of teachers for teaching mathematics with appropriate technologies. The primary goal is for mathematics teachers to integrate technology such that it becomes an extension of self (Galbraith, 2006) where it becomes an integral part of students' mathematical processing and transparent in the learning process. Attention to beliefs, pedagogical skills, lesson planning, PCK, and equity are important elements in mathematics content courses, preservice mathematics teacher education courses, practica and student teaching experiences, and inservice professional development in ways that ensure mathematics teachers are prepared for teaching.

In January 2006, the Association of Mathematics Teacher Educators (AMTE) released a technology position statement titled *Preparing Teachers to Use Technology to Enhance the Learning of Mathematics*. As identified in this position statement, mathematics teachers need knowledge for integrating appropriate technologies in the teaching of K-12 mathematics, knowledge that is actualized through:

- a deep, flexible, and connected conceptual understanding of K-12 mathematics that acknowledges the impact of technology on what content should be taught;
- a research-based understanding of how students learn mathematics and the impact technology can have on learning;
- a strong pedagogical knowledge-base related to the effective use of technology to improve mathematics teaching and learning; and
- appropriate experiences during their teacher preparation program in the use of a variety of technological tools to enhance their own learning of mathematics and the mathematical learning of others.
(AMTE, 2006, p. 1)

The expression *technological pedagogical content knowledge* (TPCK) describes this body of knowledge teachers need for teaching mathematics with appropriate digital technologies. TPCK is conceived as the interconnection and intersection of content, pedagogy (teaching and student learning), and technology (Margerum-Leys & Marx, 2002; Mishra, & Koehler, 2006; Niess, 2005). Viewed as more than a set of multiple domains of knowledge and skills, TPCK is a way of strategically thinking within these multiple domains of knowledge (Shavelson, Ruiz-Primo, Li, & Ayala, 2003). For mathematics, TPCK includes thinking that encompasses knowing when, where, and how to incorporate appropriate technologies for teaching and learning mathematics. In essence, TPCK involves **planning, organizing, critiquing,** and **abstracting** for specific mathematics content, student needs, and classroom situations while concurrently considering the multitude of 21st century technologies to support students' learning. Succinctly, AMTE's (2006) position is that:

> Mathematics teacher preparation programs must ensure that all mathematics teachers and teacher candidates have opportunities to acquire the knowledge and experiences needed to incorporate technology in the context of teaching and learning mathematics. (p. 1)

Consider four questions for thinking about this challenge: In supporting participant's development of TPCK, what is the role of the following: (1) preservice mathematics content courses; (2) preservice mathematics teacher education courses; (3) preservice teacher education practica and student teaching; and (4) inservice teacher professional development experiences?

One way to view levels of technology use in the classroom employs a taxonomy proposed by Galbraith (2006):

1. *Technology as Master* depicts students limited in their technology and mathematical knowledge.
2. *Technology as Servant* features the particular technology as a timesaving mechanical aid.
3. *Technology as Partner* shows students using the technology to increase their power to investigate, explore, and communicate with peers.
4. *Technology as an Extension of Self* is the highest level of functioning where the technology becomes an integral part of students' mathematical processing, that is, technology becomes transparent in the learning process.

These technology levels are helpful to consider as educators plan their teacher education programs to provide articulated and appropriate technology instruction and experiences for preparing elementary and secondary preservice mathematics teachers (PSTs). This preparation is particularly challenging if PSTs have not experienced TPCK in their own learning of mathematics as students. Taylor and Ronau (2006) found that not all mathematics methods courses have assignments or activities that include technology. Yet, no one course can meet the goal of developing TPCK because the skills needed to help PSTs learn to use technology as an extension of self (Galbraith, 2006) are advanced, requiring years of preparation.

Redesign of Programs for Developing TPCK

Questions abound when redesigning mathematics teacher preparation programs to prepare teachers in developing TPCK. The International Society for Technology in Education (ISTE) created the National Educational Technology Standards for Teachers (NETS•T) to shape the knowledge teachers need for teaching with technologies (ISTE, 2008). The NETS•T standards describe five areas for programs to help teachers develop TPCK: facilitate and inspire student learning and creativity; design and develop digital-age learning experiences and assessments; model digital-age work and learning; promote and model digital citizenship and responsibility; engage in professional growth and leadership (ISTE, 2008). These Standards emphasize the importance of technology in teaching and learning mathematics, positing that technology enables students to:

- visualize and experience mathematics in ways that were previously impossible,
- engage in real world problem solving,
- perform quick and complex computations, and
- create representations of their own learning.

Technologies such as the Internet, calculators, simulation and spreadsheet software, videoconferencing, and simulators hold untapped potential to enhance, modify, and connect mathematics education to applications grounded in real-world problems.

The primary issue for teacher educators is to determine which goals are most appropriate for the development of TPCK for mathematics teaching and how these goals should be articulated throughout mathematics education programs. Taylor and Ronau (2006) found that goals and activities in mathematics methods courses vary significantly. Forty-six mathematics teacher educators from 39 institutions developed six goals for mathematics methods courses, which, upon reflection, reach beyond methods courses and address the breadth of preparing mathematics teachers. Thus, these goals for mathematics teacher preparation provide a framework for integrating technology to develop TPCK:

1. *Beliefs*: Analyze and purposefully transform/build upon prospective teachers' beliefs and dispositions about what mathematics is and what it means to learn, do, and teach mathematics.
2. *Student mathematical thinking*: Engage prospective teachers in the examination of student work (i.e., listen to, look at, and reflect upon) so that they can make informed instructional decisions.
3. *Lesson planning/implementation*: Develop prospective teachers' skills in designing and implementing lessons that engage students in meaningful learning (tasks, sequence, discourse, and questioning).
4. *Reflection*: Develop prospective teachers' skills as reflective practitioners who analyze their practice from the perspective of a teacher, a researcher, a learner, and from the perspective of what and how they see students learn.
5. *Pedagogical content knowledge*: Deepen and connect mathematical content knowledge of prospective teachers with knowledge of pedagogy for teaching mathematics.
6. *Equity*: Enhance prospective teachers' ability to understand and engage in the enactment of equity and access to quality mathematics for students, parents and communities (including attention to policy). (Ronau & Taylor, 2007)

Advanced TPCK skills are required to reach Galbraith's (2006) highest level of independence and sophistication where technology becomes transparent to the learning and teaching of mathematics. The following sections address these goals and technology proficiency levels in four key components of mathematics teacher preparation: 1) preservice mathematics content courses, 2) preservice mathematics teacher education courses, 3) preservice teacher education practica and student teaching, and 4) inservice teacher professional development experiences. Although each of these areas has unique opportunities and challenges for the development of TPCK, none can or should shoulder the entire task for enhancing TPCK.

Preservice Mathematics Content Courses

Content courses for prospective mathematics teachers are critical in guiding them to examine their beliefs about what mathematics is and what it means to learn, do, and teach mathematics (Goal 1) as well as to examine their own mathematical thinking and learning (Goals 2, 3, and 4) in order to become better learners of mathematics themselves. These courses also provide the foundation for pedagogical content knowledge (PCK; Goal 5). Mathematics content courses for teachers should be taught in ways that are aligned with similar goals in other courses in the teacher preparation program. Developing TPCK requires mathematics learning experiences with technologies that differ from paper-pencil experiences. For example, calculators are widely used in schools to graph polynomial functions and explore their intersections; yet, mathematics curricula often only highlight symbolic manipulations when studying this content. Although mathematics content courses for elementary teachers tend to focus on helping them develop a conceptual understanding of the content, often through the use of a variety of manipulatives, these classes often fail to emphasize key mathematical understandings, such as variation, and integration of appropriate digital technologies. Too often, PSTs are not engaged in learning mathematics in ways that model appropriate learning experiences with digital technologies or in ways that help them develop the metacognitive strategies to master their own learning (Goos, 2006; Mishra & Koehler, 2007; Mitchell, Kelleher, & Saundry, 2007; Moreno-Armella, Hegedus, & Kaput, 2008; Nanjappa & Grant, 2004). Based on AMTE's technology position statement (2006), mathematics content courses should provide prospective teachers opportunities: 1) to gain experiences about the impact of learning mathematics and what mathematics can be learned with appropriate technologies, and 2) that enable them to experience the impact technology can have in learning mathematics.

PSTs' development of TPCK depends on technology experiences in courses that introduce the technology in mathematical contexts, address worthwhile mathematics with appropriate pedagogies, take advantage of specialized capabilities of various technologies, make connections, and incorporate multiple representations (Garofalo, Drier, Harper, Timmerman, & Shockey, 2000). In essence, appropriate uses of technology change the nature of learning and teaching about these mathematical concepts.

Although PSTs have increased experiences with various technologies, they have relatively few experiences exploring mathematical ideas with technologies (Bai & Ertmer, 2008; Bullock, 2004; Goos, 2005; Heid & Blume, 2008, Teo, Lee & Chai, 2007) They have experienced traditional classroom mathematics instruction, such as memorizing the Pythagorean theorem, applying it to calculate an unknown length for a given right triangle, and then practicing this process in similar examples. The capabilities of appropriate digital technologies provide a rich context for engaging students in delving deeper into the Pythagorean theorem by exploring the relationship between the areas of the squares created from each side of the right triangle. Advantages of dynamic geometry software include providing preservice teachers with experiences not easily accomplished with paper and pencil. They can construct a right triangle and the squares on each side of the triangle, measure the area of each square, and make conjectures about the relationship among these areas. Preservice teachers can test conjectures, taking advantage of the software to transform the triangle to create and analyze countless examples. They can extend this investigation by discussing if geometric shapes other than squares can be used to show the Pythagorean relationship. Ultimately, preservice teachers discover that the Pythagorean relationship is true as long as similar shapes are used, a realization that is rarely achieved through paper and pencil pedagogies. PSTs need such experiences and opportunities to learn mathematics with appropriate technologies in the process of developing their own TPCK.

These experiences alone are not sufficient. PSTs need to be knowledgeable about the range of tools that exist for particular mathematical tasks. For instance, the concept of slope can be discussed in a problem-solving context using spreadsheets, graphing calculators, dynamic geometry software, and interactive websites. After active engagement in constructing or enhancing their mathematical content knowledge through explorations that model ways of teaching with technology, PSTs can be asked to reflect on important questions such as:

- What mathematical content was investigated in this lesson?
- How could you use the technology to assess students' prior knowledge about the content?
- How did integrating the technology affect ways of teaching the content?
- What are the affordances and constraints of integrating the technology for this content?
- How can integrating the technology enhance or hinder your future students' understanding of the content?
- What types of questions enhance and extend students' learning of mathematics? and
- What other technologies are appropriate in teaching this content?

Such experiences integrate their thinking about the technology, the mathematics, and teaching and learning mathematics – important experiences for framing their TPCK.

Preservice Mathematics Teacher Education Courses

Preservice mathematics teacher education courses play a key role in developing TPCK for preservice mathematics teachers because such courses help prospective mathematics teachers examine their beliefs about what it means to learn, do, and teach mathematics (Goal 1) along with their students' mathematical thinking and learning (Goals 2, 3, and 6) as they develop lessons with experiences to meet their students' mathematical needs. In addition, these courses focus on developing teachers' PCK (Goal 5). Mathematics teacher education instructors report that PSTs' attitudes toward and beliefs about mathematics are not typically aligned with needs of today's classrooms; K-4 teachers exhibit high levels of mathematics anxiety (Gresham, 2008; Hoz & Weisman, 2008; Wilkins, 2008). Issues of equity arise through reports of achievement gaps associated with race or socio-economic status (Flores, 2007; Lim, 2008). Taylor and Ronau (2006) report that the development of PCK is most often cited in individual mathematics methods syllabi. The second most cited goal is lesson planning. All six of the mathematics teacher preparation goals are present in many of the syllabi, but few syllabi contain significant goals related to technology. Thus, mathematics teacher education courses will face increasing pressure to include more attention to TPCK.

Expanding PSTs' experiences in ways that address TPCK as well as these goals may involve integrating mathematics teacher education courses with mathematics content courses as Kosheleva, Rusch, and Ioudina (2006) have done. Their use of Tablet PCs facilitated significant improvement in preservice teachers' learning of the mathematical content knowledge through exploration and utilization of technology in mathematics practice teaching. In their study, the mathematics methods and mathematics content courses were team taught by faculty from Education and Science. The PSTs had opportunities to observe and analyze students' mathematical thinking in actual classes held in area schools. PSTs created innovative mathematics activities and lessons, practiced their lessons during the methods class, and taught the lessons in actual grade-level classrooms under faculty supervision. The availability and integration of the technology changed how PSTs approached the investigations and their attitudes toward mathematics.

Another important aspect of PSTs' teacher education preparation is the extent to which they can explore diverse approaches when using technology. Makar and Confrey (2006) identified three categories of epistemological approaches present in students who were conducting investigations using dynamic statistical software: *Wonderers*, *Wanderers*, and *Answerers*. The *Wonderers* approached investigations as expected by the creators of the software, i.e., testing conjectures, arriving at conclusions after their explorations. The *Wanderers* were more involved in "free" explorations while observing what relationships arise from the data, and developed more conclusions than *Wonderers* during the same period of time. *Answerers* were the most focused group – they took a shorter time period to complete the work and were more efficient. They looked for evidence from the data that allowed them to draw conclusions necessary for correct responses. This study suggests that mathematics teacher education courses should provide technology-enhanced learning environments so that PSTs have opportunities to identify diverse learners and analyze their thinking through their use of technological tools in mathematical investigations. Video cases of students' work (see for example those of Hollebrands and Lee) provide valuable experiences for PSTs (and practicing teachers) to analyze students' mathematical knowledge (Goal 2) to guide them in becoming *Answerers* (Borko, Jacobs, Eiteljorg, & Pittman, 2006; Hollebrands, Wilson, & Lee, 2007).

Preservice Teacher Education Practica and Student Teaching

Many mathematics programs require preservice teachers to design and teach lessons that integrate technology in order to address all six mathematics teacher preparation goals directed toward developing TPCK. Access issues limit the instruction to commonly found classroom technologies for these lessons, including calculators, applets, dynamic geometry software, and spreadsheets. Methods and pedagogy courses taken prior to student teaching can engage PSTs in thinking about various instructional strategies and classroom management concerns in a technology-enhanced classroom. Some teacher education programs introduce PSTs to standards-based curriculum materials in which appropriate technologies are integrated in the instruction. For example, Niess (2008) developed a set of spreadsheet activities integrated with eight (8) major algebraic topics typically considered in the first year algebra course during the middle grades; the activities were built around the *Connected Mathematics* curriculum series to extend topics and problems that engage students in learning to use the spreadsheet as a mathematics tool. Mathematics methods and pedagogy courses can engage PSTs in investigating these examples of well-planned activities to initiate discussions about what is possible and what may be missing in the lessons because of the technology. Some teacher education programs provide microteaching settings where the PSTs test their ideas for integrating the technology. Microteaching, accompanied by carefully designed plans, can lead to PSTs who are better prepared for teaching technology-enhanced lessons in their student teaching (Niess, 2005).

PSTs apprentice with expert teachers during student teaching. Cooperating teachers work one-on-one with student teachers, often sharing 'wisdom' from their extensive teaching experiences. Because many cooperating teachers have not developed their own TPCK, they are reluctant to become involved in experiences using technology (Suharwoto, 2006). Cooperating teachers may question what their students are learning when they use the technologies. Are they learning to use the technology as a crutch rather than a learning tool? Is the time spent with the technology at the expense of students learning more mathematics? Cooperating teachers are often novices with respect to their development of TPCK and yet their task is to mentor the student teacher in teaching with the technology. Alternatively, their PSTs are novices in teaching and learning and at an initial stage of development of PCK but may have more experiences than their cooperating teachers with the technologies in a mathematical context.

The communities-of-practice literature suggests a model for enhancing the field experiences for student teachers. PSTs have found great value in student teaching experiences where all members of the community (PST, cooperating teacher, university personnel) benefit from increased professional communication (Bullough et al., 2003). This notion suggests a community-of-practice around the development of TPCK so that the student teacher and the cooperating teacher engage in a collaborative partnership during student teaching. Together the student teacher and the cooperating teacher can plan to integrate technology in a specific mathematics unit. The PST could be responsible for the technology skills students need during the unit while the cooperating teacher could be responsible for guiding the specific instructional strategies focused around the needs of all students. Thus, the student teacher and cooperating teacher work collaboratively, sharing their respective specific expertise while focusing on the integration of the mathematics and the technology. Each partner teaches the lessons in similar classes; each observes the work of the other; and, after each lesson, they can discuss the results and strategize changes and embellishments to meet specific students' ways of thinking. In the process, they engage in collaboration around **planning, organizing, critiquing,** and **abstracting** for specific mathematics content, student needs, and classroom situations while concurrently considering appropriate technologies to support students' learning. Thus, both the preservice teacher and the cooperating teacher extend their TPCK (Harrington, 2008).

Inservice Teacher Professional Development Experiences

Many inservice mathematics teachers have limited experiences using technologies as mathematics learning tools (Bullock, 2004; Goos, 2005; Heid & Blume, 2008; Niess, Sadri, & Lee, 2007). Yet, they are expected to design lessons in which students learn about the technologies as well as use the technologies to explore mathematics. Their professional development must include TPCK development that focuses on the knowledge required for scaffolding students' learning of mathematics as they also learn about the technology. This work must challenge them to reconsider their mathematics content and develop their knowledge of the technology as well as provide evidence of technology's impact on the learning of mathematics. The process of learning to teach - a "constructive and iterative" process - must engage the teachers in "events on the basis of existing knowledge, beliefs, and dispositions" (Borko & Putnam, 1996, p. 674).

In studying teacher decision-making about technology, Weinhold (2004) asked groups of teachers to collaborate on a document (Zawojewski, Chamberlin, Hjalmarson, & Lewis, 2008) to guide decisions about when and how to use graphing calculators to teach mathematics. Weinhold found that after introduction to Zbiek's (2002) categories of what students do with technology, the inservice teachers using standards-based curricula were more likely to incorporate the broad range of student activities into their document than they did before the introduction. Because technology is an integral part of many standards-based curricula, these teachers created more examples for thinking about technology, and were better able to understand and assimilate Zbiek's categories. Teachers using more traditional curricula were more likely to focus on knowledge they believed students needed before using calculators (cf. Fleener, 1995).

Shreiter and Ammon (1989) argue that teachers' adaptation of new instructional practices is a process of assimilation and accommodation resulting in changes in their thinking. Teachers need experiences to investigate, think, plan, practice, and reflect. They need active learning - not only about the technology but also about teaching and learning mathematics with the technology. Professional development programs must provide follow-up support to assist teachers in implementing their instructional plans and in adopting new curriculum and instructional strategies to guide student learning of mathematics with technologies (Feist, 2003).

The six mathematics teacher preparation goals from this chapter framed a comprehensive professional development program directed at the development of TPCK for teaching mathematics with spreadsheets. Niess, Sadri, and Lee (2007) held a summer workshop focused on using spreadsheets for teaching and learning mathematics followed by support for teaching mathematics with spreadsheets during the following school year. The teachers learned about spreadsheets in units on variables, covariation, and multiple representations; this experience emphasized the value of scaffolding to build knowledge of spreadsheet capabilities while emphasizing spreadsheets as tools for learning mathematics. The teachers collaborated to design lessons and teach a summer class of middle school students. After each day of instruction, the teachers reflected on and revised the lessons in order to implement them in their own classes during the school year. Yearlong follow-up support staff reported that the teachers improved their levels of TPCK for teaching mathematics with spreadsheets. Niess, Sadri, and Lee (2007) also proposed multiple levels of TPCK that could guide mathematics teacher educators in assessing inservice teachers' changing TPCK.

Conclusion

With the recognition of the importance of developing TPCK for teaching mathematics with appropriate technologies, teacher preparation programs (both preservice and inservice) are challenged to rethink their approaches for preparing mathematics teachers to teach with technology. A set of carefully agreed upon goals (such as the mathematics teacher preparation goals we used in this chapter) must frame content and mathematics teacher education courses, field placement and student teaching, and inservice programs. To teach with technologies in ways that appear transparent and natural – an extension of self – requires focused strategic efforts to reach Galbraith's (2006) highest level of sophistication. Attention to beliefs, pedagogical skills, lesson planning, PCK, and equity are important elements to ensure PSTs are ready for their future classrooms. PSTs must be prepared to change, to become better learners themselves, and to plan for extending their TPCK. A campus-wide approach that fully integrates technology with a variety of teaching approaches in all the prospective teachers' courses seems the most reasonable. However, this challenge calls for more research and evaluation of how both inservice and preservice mathematics teachers' TPCK is developed and enhanced in ways that support K-12 students to learn important mathematics in the 21st century.

References

Association of Mathematics Teacher Educators. (2006). *Preparing teachers to use technology to enhance the learning of mathematics*. Retrieved May 20, 2007, from http://www.amte.net/

Bai, H. & Ertmer, P. (2008). Teacher educators' beliefs and technology uses as predictors of preservice teachers' beliefs and technology attitudes. *Journal of Technology and Teacher Education, 16* (1), 93-112.

Borko, H., Jacobs, J., Eiteljorg, E. , & Pittman, M. E. (2006). Video as a tool for fostering productive discussions in mathematics professional development. *Teaching and Teacher Education, 24*(2), 417-436.

Borko, H., & Putnam, T. (1996). Learning to teach. In D. C. Berliner & R. C. Calfee (Eds.), *Handbook of educational psychology* (pp. 673-708). New York: Simon & Schuster Macmillan.

Bullock, D. (2004). Moving from theory to practice: An examination of the factors that preservice teachers encounter as the attempt to gain experience teaching with technology during field placement experiences. *Journal of Technology and Teacher Education, 12*(2), 211-237.

Bullough, R. V., Young, J., Birrell, J., Clark, D. C., Egan, M. W., Erickson, L., et al. (2003). Teaching with a peer: A comparison of two models of student teaching. *Teaching and Teacher Education, 19*, 57-73.

Feist, L. (2003, June 1). Removing barriers to professional development. *T.H.E. Journal Online Technological Horizons in Education*. Retrieved September 27, 2008, from http://www.thejournal.com/articles/16391

Fleener, M. J. (1995). A survey of mathematics teachers' attitudes about calculators: The impact of philosophical orientation. *The Journal of Computers in Mathematics and Science Teaching, 14*(4), 481-498.

Flores, A. (2007). Examining disparities in mathematics education: Achievement gap or opportunity gap? *The High School Journal, 91*(1), 29-42.

Galbraith, P. (2006). Students, mathematics, and technology: Assessing the present - challenging the future. *International Journal of Mathematical Education in Science and Technology, 37*(3), 277-290.

Garofalo, J., Drier, H., Harper, S., Timmerman, M. A., & Shockey, T. (2000). Promoting appropriate uses of technology in mathematics teacher preparation. *Contemporary Issues in Technology and Teacher Education* [online serial], *1* (1). Available: http://www.citejournal.org/vol1/iss1/currentissues/mathematics/article1.htm

Goos, M. (2005). A sociocultural analysis of the development of preservice and beginning teachers' pedagogical identities as users of technology. *Journal of Mathematics Teacher Education*, 8(1), 35-59.

Goos, M. (2006, December). Understanding technology integration in secondary mathematics: Theorizing the role of the teacher. Paper presented at the Digital Technologies and Mathematics Teaching and Learning: The 17th

ICMI Study Conference, Hanoi. [Retrieved September 26, 2008 from http://espace.uq.edu.au/eserv/UQ:104162/UQ_AV_104162.pdf].

Gresham, G. (2008). Mathematics anxiety and mathematics teacher efficacy in elementary preservice teachers. *Teaching Education, 19*(3), 171-184.

Harrington, R. A. (2008). *The development of preservice teachers' technology specific pedagogy*. Unpublished doctoral dissertation, Oregon State University, Oregon.

Heid, M. K., & Blume, G. W. (2008). Technology and the teaching and learning of mathematics. In M. K. Heid & G. W. Blume (Eds.), *Research on the teaching and learning of mathematics: Synthesis and perspectives,* 419-431. Charlotte, NC: Information Age Publishers.

Hollebrands, K., Wilson, P. H., & Lee, H. S. (2007, October 25). *Prospective teachers' use of a videocase to examine students' work when solving mathematical tasks using technology*. Paper presented at the annual meeting of the psychology of Mathematics Education - North American Chapter. Lake Tahoe, CA/NV. Retrieved September 27, 2008 from http://www.allacademic.com//meta/p_mla_apa_research_citation/1/8/8/5/6/pages188561/p188561-1.php.

Hoz, R., & Weizman, G. (2008). A revised theorization of the relationship between teachers' conceptions of mathematics and its teaching. *International Journal of Mathematical Education in Science and Technology, 39*(7), 905-924.

International Society for Technology in Education. *National educational technology standards (NETS), second edition.* Retrieved September 27, 2008, from http://www.iste.org/AM/Template.cfm?Section=NETS.

Kosheleva, O., Rusch, A., & Ioudina, V. (2006). Analysis of effects of Tablet PC technology in mathematical education of future teachers. *Electronic Proceedings of the Seventeenth ICMI Study*. Hanoi, Vietnam: ICMI.

Lim, J. H. (2008). Double jeopardy: The compounding effects of class and race in school mathematics. *Equity & Excellence in Education, 41*(1), 81-97.

Makar, K., & Confrey, J. (2006). Dynamic statistical software: How are learners using it to conduct data-based investigations? *Electronic Proceedings of the Seventeenth ICMI Study*. Hanoi, Vietnam: ICMI.

Margerum-Leys, J., & Marx, R. W. (2002). Teacher knowledge of educational technology: A study of student teacher/mentor teacher pairs. *Journal of Educational Computing Research, 26*(4), 427-462.

Mishra, P., & Koehler, M. J. (2006). Technological pedagogical content knowledge: A framework for integrating technology in teacher knowledge. *Teachers College Record, 108*(6), 1017-1054.

Mishra, P. & Koehler, M. (2007). Technological pedagogical content knowledge (TPCK): Confronting the wicked problems of teaching with technology. In C. Crawford et al. (Eds.), *Proceedings of Society for Information Technology and Teacher Education International Conference 2007* (pp. 2214-2226). Chesapeake, VA: Association for the Advancement of Computing in Education.

Mitchell, J., Kelleher, H., & Saundry, C. (2007). Learning by design: A multimedia mathematics project in a teacher education program. In L. F. Darling, G. Erickson, & A. Clark (Eds.), *Collective improvisation in a teacher education community* (101-118). New York: Springer.

Moreno-Armella, L., Hegedus, S. J., & Kaput, J. J. (2008). From static to dynamic mathematics: Historical and representational perspectives. *Educational Studies in Mathematics, 68*(2),99-111.

Nanjappa, A., & Grant, M. (2004). Constructing on constructivism: The role of technology. *Electronic Journal for the Integration of Technology in Education, 2*(1). Retrieved September 28, 2008 from http://ejite.isu.edu/Volume2No1/Nanjappa.htm.

Niess, M. L. (2005). Preparing teachers to teach science and mathematics with technology: Developing a technology pedagogical content knowledge. *Teaching and Teacher Education, 21*(5), 509-523.

Niess, M. L. (2008). *Integrating Spreadsheets in Middle School Mathematics Algebra: Complete Curricular Kit.* (September, 2008). Retrieved October 1, 2008 from http://eusesconsortium.org/edu/education.php

Niess, M. L., Sadri, P., & Lee, K. (2007, April). *Dynamic spreadsheets as learning technology tools: Developing teachers' technology pedagogical content knowledge (TPCK).* Paper presented at the meeting of the American Educational Research Association Annual Conference, Chicago, IL.

Ronau, R. N., & Taylor, P. M. (2007, January 25). Analysis of mathematics methods courses. Presentation at the annual meeting of the Association of Mathematics Teacher Educators (AMTE), Irvine, CA.

Shavelson, R., Ruiz-Primo, A., Li, M., & Ayala, C. (2003, August). *Evaluating new approaches to assessing learning* (CSE Report 604). Los Angeles, CA: University of California, National Center for Research on Evaluation.

Shreiter, B., & Ammon, P. (1989). *Teachers' thinking and their use of reading contracts.* Paper presented at the meeting of the American Educational Research Association, San Francisco, CA.

Suharwoto, G. (2006). Secondary mathematics preservice teachers' development of technology pedagogical content knowledge in subject-specific, technology integrated teacher preparation program. Unpublished doctoral dissertation, Oregon State University, Oregon.

Taylor, P. M., & Ronau, R. (2006). Syllabus study: A structured look at mathematics methods courses. *AMTE Connections, 16*(1), 12-15.

Teo, T., Lee, C. B., & Chai, C. S. (2007). Understanding preservice teachers' computer attitudes: Applying and extending the technology acceptance model. *Journal of Computer Assisted Learning, 24*(2), 128-143.

Weinhold, M. (2004). Constructing an understanding of "appropriate use" of graphing calculators in the context of collegial inquiry. *Wisconsin Teacher of Mathematics, 55*(1), 7-13.

Wilkins, J. L. M. (2008). The relationship among elementary teachers' content knowledge, attitudes, beliefs, and practices. *Journal of Mathematics Teacher Education, 11*(4), 139-164.

Zawojewski, J., Chamberlin, M., Hjalmarson, M., & Lewis, C. (2008). Developing design studies in mathematics education professional development: Studying teachers' interpretive systems. In A. E. Kelly, R. A. Lesh, & J. Baek (Eds.), *Handbook of design research methods in education: Innovations in science, technology, engineering, and mathematics learning and teaching*. London: Routledge.
Zbiek, R. M. (2002). *A two-tiered category perspective to describe purposes of mathematics technology use.* Paper presented at the annual meeting of the National Council of Teachers of Mathematics, Las Vegas, NV.

Margaret L. Niess is a Professor Emeritus of Mathematics Education in Science and Mathematics Education at Oregon State University. Her research interests include the knowledge teachers need for teaching with technology (TPCK), preparation of preservice and inservice teachers to teach with technology, integrating spreadsheets as a mathematics learning tool.

Robert N. Ronau is a Professor of Mathematics Education and associate dean for research at the University of Louisville. His research interests are focused on mathematics knowledge for teaching, assessment, technology for teaching mathematics, and mathematics learning.

Shannon Driskell is an Assistant Professor of Mathematics Education at the University of Dayton in Dayton, Ohio. Her research interests include the appropriate use of technology to teach K–12 mathematics, the content knowledge of prospective mathematics teachers, and the teaching and learning of geometry.

Olga Kosheleva is an Assistant Professor of Mathematics Education at the University of Texas at El Paso. Her research interests include preservice and inservice mathematics teacher education, effective embedding of technology into teaching and learning, research in mathematics and applied mathematics.

David Pugalee is Professor of Education and Research Associate in the Center for Mathematics, Science & Technology Education at the University of North Carolina Charlotte. His research interest is mathematical literacy: the relationship between language and mathematics learning.

Marcia Weller Weinhold is an Assistant Professor of Mathematics Education at Purdue University Calumet. Her research interests include issues teachers consider when deciding whether to use technology in teaching mathematics, and activities that help preservice and inservice teachers use technology appropriately.

Reed, M. and Mathews, S.
AMTE Monograph 5
Inquiry into Mathematics Teacher Education
©2008, pp. 157-166

14

Scholarship for Mathematics Educators: How Does This Count for Promotion and Tenure?

Michelle K. Reed
Susann M. Mathews
Wright State University

Within some departments of mathematics that house mathematics educators, there is a chasm between the mathematicians and the mathematics educators regarding the definition of scholarship. This becomes particularly apparent when making promotion or tenure decisions and during annual evaluations. We suggest a framework for developing a dialogue between mathematicians and mathematics educators to broaden the traditional definition of scholarship. We first provide the historical context of scholarship within universities, followed by Boyer's broadened definition of scholarship. We discuss professional practice, and organize it according to Boyer's types of scholarship and to recommendations from the Joint Policy Board of Mathematics.

> The specialization of science is an inevitable accompaniment of progress; yet it is full of dangers, and it is cruelly wasteful, since so much that is beautiful and enlightening is cut off from most of the world. Thus it is proper to the role of the scientist that he not merely find the truth and communicate it to his fellows, but that he teach, that he try to bring the most honest and most intelligible account of new knowledge to all who will try to learn.
>
> Robert Oppenheimer, Physicist, 1954 (as cited in Boyer, 1990)

Tenure decisions are not taken lightly by departments or by candidates. At some institutions of higher education, and particularly within mathematics departments, there seems to be a chasm between mathematicians and mathematics educators regarding the definition of scholarship. As mathematics educators who are housed in a mathematics department, we and our mathematician colleagues have struggled with this definition. We suppose this struggle occurs in many, if not every, mathematics department that houses mathematics educators. Our purpose in the chapter is to initiate a dialogue regarding the weighting or counting of professional practice in promotion and tenure decisions in mathematics departments. Our intention is to provide a framework so the mathematics education community may define the idea of scholarship for their mathematician colleagues to begin a dialogue within the department.

We begin this chapter with the historical context of universities in the United States, including the corresponding changing focus on scholarship throughout this historical development. We then define and provide examples of Boyer's (1990) broadened definition of scholarship. We then discuss the obstacles within academia and within mathematics departments, in working with this broader definition of scholarship along side recommendations from the Joint Policy Board for Mathematics [JPBM] (1995). We conclude by looking at professional practice and the advantages and disadvantages of an example of one mathematics department's bylaws when applying these ideas to mathematics educators.

Historical Context

Boyer's (1990) seminal book, *Scholarship Reconsidered*, presents a cogent history of the growth of university faculty roles in the United States. The fundamental role of faculty has developed through three distinct stages, from colonial times to the late twentieth century. In the seventeenth century, when the first U.S. universities were created for the economically elite, the role of the faculty was to focus on their students. Professors mentored their students both intellectually and morally to prepare them for their future roles as civic and religious leaders. Thus, teaching lay at the heart of a professor's career.

This focus lasted well into the nineteenth century. However, with the opening of Rensselaer Polytechnic Institute in 1836 as one of the first technical schools and the Land Grant College Act of 1862 with its subsequent land-grant universities created to research agriculture, institutes of higher education added service to the nation as a primary component of faculty's responsibilities. Faculty were expected to use their expertise toward the success of the agricultural and mechanical revolutions. They were still expected to prepare the next generation of leaders but now they were expected to guide them toward leadership in agriculture and manufacturing. Critics saw this added responsibility as diluting academic standards, both because of the focus on agriculture and because non-elite young people now were attending colleges.

In 1876 Daniel Coit Gilman introduced the German paradigm of education into the U.S. when he founded Johns Hopkins University. The German model held that scholarship was best attained through research and experimentation. The ascendance of basic research to the fore of scholarship during the mid-nineteenth century began the change toward an environment in which research productivity became the norm for promotion and tenure decisions in universities throughout the U.S. (Boyer, 1990).

Broadened Definition of Scholarship

The Carnegie Foundation's national survey of college faculty (Glassick, 2000) found that the majority of faculty felt it very difficult to achieve tenure and promotion in their department if they did not publish. Furthermore, the type of scholarship that is almost exclusively valued and rewarded is that of traditional research (scholarship of discovery) published in refereed journals and high-quality books (Braxton, Luckey, & Helland, 2006). A rough hierarchy exists that rates theoretical topics at the top and pedagogical issues at the bottom (Gebhardt, 1995). Many mathematics faculty do not believe that writing about pedagogy is legitimate scholarship, even if it is the scholarship of discovery. Thus, as we work toward expanding the definition of scholarship, we contend that we need to expand the inclusiveness of the scholarship of discovery to include writing about mathematics education as well as work to broaden the range of acceptable writing to include textbooks and articles about teaching.

With the publication of *Scholarship Reconsidered* (1990), Boyer broadened the definition of scholarship beyond that of basic research, or the scholarship of discovery, and provided a common vocabulary to debate issues of scholarship (Glassick, 2000). Although the scholarship of discovery was still considered the heart of academic life, Boyer added the scholarship of integration, of application, and of teaching. Mathematics educators and many in academia have welcomed this extension of the definition of scholarship.

Although the scholarship of discovery is that of searching for new knowledge, the scholarship of integration is that of making connections across disciplines. It includes work that interprets original research and sets it into a larger intellectual context among other areas of study. It can illuminate data and results so that non-specialists can understand the research. This type of integration happens when mathematicians and mathematics educators work together to investigate children's thinking or other issues of pedagogy (Ball, Ferinni-Mundy, et al., 2005). A well-known instance of this integration is the work of Deborah Ball and Hyman Bass (Ball & Bass, 2003; Ball, Hill, & Bass, 2005).

The scholarship of application relates theory and research to practical use in real life. In mathematics education, we participate in this type of scholarship when we responsibly apply our professional knowledge to our work with K-12 schools. We conduct professional practice tied directly to our areas of expertise.

One specialty of a mathematics educator in our mathematics department is writing articles about research relevant to teachers. In this scholarship of application, she helps teachers interpret research findings by summarizing results across several studies on one topic or by explaining the type of studies done to help teachers interpret for themselves the results of other research articles. She has published both peer-reviewed and invited papers in this area.

The scholarship of teaching occurs when teaching is more than just transmitting knowledge and goes on to "transform and extend it as well" (Boyer, 1990, p. 24). There is a widespread view that teaching is just holding office hours, teaching class, and grading student work (and that grading mathematics work must be quick and easy because one simply has to grade the accuracy of students' results). However, as scholarship, teaching depends upon the professor being steeped in the knowledge of his/her field, inspiring future scholars in the discipline, and "building bridges between the teacher's understanding and the student's learning" (Boyer, 1990, p.23).

Definitions of the scholarship of teaching are wide in range. Lee Shulman expanded the focus of this scholarship by using the phrase "scholarship of teaching and learning" (SOTL) (Bender, 2005). Hutchings and Shulman (1999) stress the difference between excellent teaching and scholarship of teaching; the latter includes a public account of the act of teaching and "involves question-asking, inquiry, and investigation, particularly around issues of student learning" (p. 13). An example of this type of scholarship would be involvement in the process of action research or of lesson study (Lewis, Perry, Hurd, & O'Connell, 2006). Other examples can be found among projects in CASTL—the Carnegie Academy for the Scholarship of Teaching and Learning (Bender, 2005).

Overcoming Hurdles

The more comfortable a faculty member is with the present system, the greater the possibility that he or she will be resistant to change in the promotion and tenure system. It was very clear from data in the National Study that many faculty, particularly in the sciences, see their departments losing resources, and consequently power, if the present reward system is modified (Diamond, 1995, p. 26).

Any discussions about changes in the reward system for faculty must begin within our departments. It is at the departmental level where the mission of the institution, the nature of the discipline, the priorities within the department and college, and individual strengths and interests first intersect (Diamond, 1999). Furthermore, if policy statements or bylaws need to be written, it is the department that will determine the requirements for promotion and tenure and for annual evaluations.

Because we are discussing the case of scholarship within mathematics departments, it is imperative that we understand mathematicians' point of view. First, we need to know if the primary priority in the department is teaching or research. If teaching is the department's main priority, then this discussion may be moot. However, assuming that we are in a mathematics department with research as the primary focus, we need to internalize the fact that research mathematicians will vote on our promotion and tenure cases. Thus, we must understand and appreciate their points of view. Mathematics research is abstract and part of the body of research that is most valued and respected (scholarship of discovery). Research mathematicians see their discipline as a pinnacle of human intellectual achievement. Mathematics research is a vast and mature field with 3,000 subfields in which approximately 75,000 items are published each year. Compared to this body of work, mathematics education research is still relatively young and is an order of magnitude smaller, at about 4,500 items per year. Further, mathematics education research is high in sociological and psychological content. Not only is mathematics research very low in sociological and psychological content, most mathematicians are uncomfortable with these areas (R. Mercer, personal communication, January 27, 2007).

Keeping these ideas in mind, we still need to work toward consensus within our mathematics departments when we develop department performance expectations. We should not try to claim that mathematics research and mathematics education research are similar nor try to compare mathematics publications with mathematics education publications. Instead, we need to define the kinds of work appropriate for faculty in mathematics and for those in mathematics education; we must agree upon evidence that is acceptable for documenting that work; we should have differences in performance expectations for faculty at different ranks; and we should identify service that is valued and rewarded (Diamond, 1999).

Because we are looking at issues of "what counts for promotion and tenure," and because traditional research and publishing may not fit well the interests and priorities of many mathematics educators, we need to work within our departments to broaden the range of valued writing. However, those immersed in basic research, such as many mathematics faculty:

- May not know of the expanded definition of scholarship;
- May know it but not agree with it;
- May agree to an extent, but not value scholarship other than traditional research; or
- May agree whole-heartedly but have little idea how to evaluate different types of scholarship.

As we work with mathematicians to develop clear guidelines for promotion and tenure, we should not only share definitions of scholarship to include integration, application, or teaching, but also as importantly, provide scholarly examples of each within mathematics education. These examples will be even more helpful if they are accompanied with appropriate documentation to demonstrate their value and scholarliness. Furthermore, as Diamond notes "it will be the responsibility of your academic unit to determine if the activity or work itself falls within the priorities of the department,…it will be your responsibility to provide substantiation of the significance and quality of your work" (2004, p. 20).

Alternative Scholarship Practices and Needed Documentation

The JPBM Committee on Professional Recognition and Rewards (1995) stated that the reward system in mathematics departments must "encompass the full array of faculty activity required to fulfill departmental and institutional missions" (p. 55). At the same time, the committee recognized that few departments have grappled with this issue. To enable departments to value diverse faculty work, the committee offered guiding principles that are consistent with the suggestions presented in this chapter concerning the scholarship of mathematics educators. It recognized that the definition of mathematics scholarship could range from the very narrow of only research leading to knowledge that can be published to the very broad of any activity that leads to increased understanding or knowledge on the part of the individual faculty. Therefore, the report concluded with the recommendation that each mathematics department should "formulate an explicit and public definition of scholarship" (p. 65). The committee drafted guidelines for scholarship to include the following:

- Research in
  - Core and applied areas
  - Mathematical techniques and application of techniques
  - Teaching and learning;
- Syntheses of existing scholarship
  - Surveys
  - Book reviews
  - Lists of open problems;
- Expositions communicating mathematics to new audiences;
- Development of courses, curricula, or instructional materials;
- Development of software for research and teaching.

Just as the JPBM recommends the expansion of scholarship for mathematicians, definitions of scholarship need to be broadened to include the professional practice of mathematics educators. Boyer (1990) notes that each item in this expanded list must satisfy the following criteria no matter the category of scholarship or field of study: (a) serious, demanding work requiring rigor traditionally associated with research activities, and (b) directly related to one's special field of knowledge. Now the question is: What specific activities constitute scholarship for mathematics educators?

Figure 1 includes a variety of professional activities commonly associated with the work of mathematics educators that could be considered to satisfy Boyer's requirements. The table is organized within Boyer's definitions of scholarship with possible sources of evidence listed in the fourth column of the table. Items for mathematics scholarship from the JPBM Committee on Professional Recognition and Rewards (1995) are included in the table for comparison purposes.

Boyer's Definitions of Scholarship (Boyer, 1990)	Scholarship for Mathematics (JPBM, 1995)	Scholarship for Mathematics Education	Evidence
Scholarship of Discovery Investigate new information • What is to be known? • What is yet to be found?	• Research in core and applied areas • Research in mathematical techniques • Research in teaching and learning	• Research on mathematics teaching and learning • Collection of data evaluating mathematics programs in K-12 and college or comparison of programs	• Publications • Books • Conference papers
Scholarship of Integration Put isolated facts into perspective • What do the findings mean? • Is it possible to interpret what's been discovered in ways that provide a larger, more comprehensive understanding?	• Syntheses of existing scholarship	• Syntheses of existing scholarship • Projects between mathematics educators and experts from other fields • Surveys • Book reviews	• Publications • Books • Conference papers and presentations • Surveys • Book reviews • List of open problems
Scholarship of Application Apply knowledge to serve both community and campus • How can knowledge be responsibly applied to problems? • How can it be helpful to people and institutions?	• Development of software for research and teaching • Applying mathematical techniques to other fields	• Work with K-12 and college teaching • Grants • Committees and commission work—statewide, nationwide—to advise leaders • Develop standards for content and teaching • Interpreting research for teachers	• Software programs • Letter from commission • Technical reports • Standards documents • Grant reports • Grant money received
Scholarship of Teaching Mastery of knowledge and presentation of information • How can knowledge best be transmitted to others and best learned?	• Expositions communicating mathematics to new audiences orally or in writing • Development of courses, curricula, or instructional materials	• Expositions communicating mathematics to new audiences • Teaching methods/content classes • Communicating mathematics teaching to teachers • Development of courses, curricula, or instructional materials • Mentor new faculty	• Publications • Conference presentations and papers • Technical communications • Multimedia materials • Curricular materials • Materials and syllabi developed for workshops • Course portfolio

Figure 1. Scholarship activities for mathematics and mathematics education

Dialogue Regarding Interpretation and Assessment of Scholarship

Criteria for judging the quality of scholarship work and the quality of the documentation provided as evidence, along with samples of documents, will be needed to further the discussion regarding what constitutes professional practice for mathematics educators (Diamond, 2004). A possible framework for interpreting and praising scholarship could be that the scholar must have been guided by clear goals, be adequately prepared, use appropriate methods, and achieve outstanding results. Then the scholar must present the work effectively and follow up with a reflective critique (Glassick, Huber, & Maeroff, 1997). We, along with our mathematics department colleagues, have yet to come to such a common understanding of the definition and judgment of scholarship.

Since our university became unionized, each department and college was required to create bylaws. The departmental bylaws include requirements for annual evaluations as well as for promotion and tenure and were approved by department members as well as the dean of the college. To create these bylaws, much discussion occurred regarding previous, but not written, practices and requirements for promoting candidates. At that time two mathematics educators were members of the department at the rank of associate professor. Both had received tenure under the unwritten rules, so precedence was set regarding awarding tenure to mathematics educators.

The obvious advantage of having bylaws is the explicit written requirements for tenure instead of allowing guesswork to rule a pre-tenured faculty member's activities. An assistant professor can steer her activities toward those listed requirements. However, some worry that bylaws may become a checklist of activities to do instead of forming a program of scholarship. Our bylaws require evidence of such a program.

During the initial discussion of scholarship, many activities were suggested and are listed as possible scholarship venues in the bylaws under the heading "Professional Practice for Mathematics Educators." These include professional development programs, assessment of these programs or of preservice programs involving the collection of data, development of curricula published in a peer reviewed publication, and field testing curricular materials. However, when the bylaws are read carefully, these items are not specifically named scholarship work for annual evaluations or for promotion and tenure and so are not necessary activities. They are, in the words heard by one of the authors many times, "activities that help your case." The items that are listed as necessary include various stages of submitted and accepted peer-reviewed publications or invited research presentations. However, these publications are not limited to research (scholarship of discovery), but may be publications that are pedagogical in nature (scholarship of application or of teaching).

Interpretation of these bylaws continues each year as tenure-track faculty members are reviewed in their progress toward promotion and/or tenure. Discussions at these meetings include various points of difference in the nature of the two disciplines with learning occurring for all present. For instance, many mathematics faculty were surprised to learn that presenting at many mathematics education conferences requires peer-reviewed acceptance of a proposal and is not an automatic right. They were also surprised to learn that when several names are listed in a presentation, that all those faculty members participate in the presentation, and that these presentations are sometimes 90 minutes long. The continuation of these discussions is vital to a growth in understanding among those of different disciplines.

Conclusion

However it begins, dialogue between faculty in the disciplines of mathematics education and mathematics must take place. This dialogue may take many forms, with varying degrees of formality. Specific examples and documents should be helpful to make the case for a broader definition of scholarship for mathematics education. Also important would be a policy statement agreed upon by a group of mathematics educators detailing many of these items. We hope that a dialogue can begin within the Association of Mathematics Teacher Educators for this purpose.

References

Ball, D. L., & Bass, H. (2003). Toward a practice-based theory of mathematical knowledge for teaching. In B. Davis & E. Simm (Eds.), *Proceedings of the 2002 Annual Meeting of the Canadian Mathematics Education Study Group* (pp. 3-14). Edmonton, AB: CMESG/ GCEDM.

Ball, D. L., Ferrini-Mundy, J., Kilpatrick, J., Milgram, R. J., Schmid, W., & Schaar, R. (2005). Communications: Reaching for a common ground in K-12 mathematics education. *Notices of the American Mathematical Society, 52*(9), 1055-1058.

Ball, D. L., Hill, H. C., & Bass, H. (2005). Knowing mathematics for teaching: Who knows mathematics well enough to teach third grade, and how can we decide? *American Educator, Fall*, 14-17, 20-22, 43-46.

Bender, E. T. (2005). CASTLs in the air: The SOTL "movement" in mid-flight. *Change, 37*(5), 40-49.

Boyer, E. L. (1990). *Scholarship reconsidered: Priorities of the professoriate.* Princeton, NJ: Carnegie Foundation for the Advancement of Teaching.

Braxton, J. M., Luckey, W. T., & Helland, P. A. (2006). Ideal and actual value patterns toward domains of scholarship in three types of colleges and universities. *New Directions for Institutional Research, 129*, 67-76.

Diamond, R. M. (1995). Disciplinary associations and the work of faculty. In J. M. Moxley & L. T. Lenker (Eds.), *The politics and processes of scholarship* (pp. 19-26). Westport, CT: Greenwood Press.

Diamond, R. M. (1999). *Aligning faculty rewards with institutional mission: Statements, policies and guidelines*. Bolton, MA: Anker Publishing Co.

Diamond, R. M. (2004). *Preparing for promotion, tenure, and annual review: A faculty guide*. Bolton, MA: Anker Publishing Company, Inc.

Gebhardt, R. C. (1995). Avoiding the "research versus teaching" trap: Expanding the criteria for evaluating scholarship. In J. M. Moxley & L. T. Lenker (Eds.), *The politics and processes of scholarship* (pp. 10-17). Westport, CT: Greenwood Press.

Glassick, C.E. (2000). Reconsidering scholarship. *Journal of Public Health Management & Practice, 6*(1), 4-9.

Glassick, C. E., Huber, M. T., & Maeroff, G. I. (1997). *Scholarship assessed: Evaluation of the professoriate.* San Francisco: Jossey-Bass.

Hutchings, P., & Shulman, L. S. (1999). The scholarship of teaching: New elaborations, new developments. *Change, 31*(5), 10-15.

Joint Policy Board for Mathematics. (1995). Recognition and rewards in the mathematical sciences. In R. M. Diamond & E. A. Bronwyn (Eds.), *The disciplines speak: Rewarding the scholarly, professional, and creative work of faculty* (pp. 55-67). Washington, DC: American Association for Higher Education.

Lewis, C., Perry, R., Hurd, J., & O'Connell, M. P. (2006). Lesson study comes of age in North America. *Phi Delta Kappan, 88*(4), 273-281.

Michelle K. Reed is an Associate Professor in the Department of Mathematics and Statistics at Wright State University. She is currently involved in researching effects of lesson study on preservice teachers' abilities and knowledge.

Susann M. Mathews is a Professor in the Department of Mathematics and Statistics at Wright State University. Her interests lie in mathematical modeling, both historically and currently, including how modeling can help students connect mathematics to other disciplines and with the world.